Professional Learning Focus Areas come to life in each chapter's Coach's Digest and Coach's Toolkit

Chapter 3

Content Knowledge and Worthwhile Tasks

Dear Coach:

Here is support for you as you work with teachers on content knowledge and worthwhile tasks.

In the COACH'S DIGEST...

Overview: Highlights about content knowledge and tasks for you to download and share with teachers.

Coaching Considerations for Professional Learning: Ideas for how to support a teacher or a group of teachers in deepening their content, as well as exploring content and worthwhile tasks for their students.

Coaching Lessons from the Field: Story from a mathematics leader sharing how she used tools from this chapter.

Connecting to the Framework: Specific ways to connect selected *Shifts* and Mathematical Practices to content and tasks.

Coaching Questions for Discussion: Menu of prompts for professional learning or one-on-one coaching about content knowledge and using worthwhile tasks.

Where to Learn More: Articles, books, and online resources for you and your teachers with more examples of content knowledge and worthwhile tasks!

In the COACH'S TOOLKIT...

Ten tools focused on content knowledge and worthwhile tasks, for professional learning or coaching cycles.

Coach's Digest

In the Coach's Digest, we begin with an overview of content knowledge and worthwhile tasks, written to teachers (and to you, the coach). As you read the Overview, the following questions might help you reflect on this topic in terms of your role as a mathematics coach:

- How might I use the Levels of Cognition to help teachers analyze tasks and instruction?
- How might I provide experiences for teachers in opening up and increasing the cognitive demand of tasks?

The key components of the Digest and Toolkit are replicated in each chapter. An overview section is followed by special coaching considerations and lessons from the field, connections to an overarching coaching framework, questions for discussion, additional resources, and practical tools for immediate use in professional learning sessions.

Coaching Considerations for Professional Learning ●·············

Hopefully, the discussion to this point has communicated the critical importance of teachers thinking deeply about the content they are teaching. Here are some ways to do this related to content knowledge.

1. ***Address myths, misinterpretations, and overgeneralizations***. Developing a shared understanding of what the goals are in developing content is critical—yet there are many sound bites that can interfere with our shared understanding of developing content. Here are a few:

    a. *Conceptual knowledge is good, and procedural knowledge is bad (drill and kill, etc.).* In fact, both are needed to ensure mathematical proficiency (NRC, 2001; Star, 2015). *Appropriate* balance and connections between them is critical. This must be our message if we are going to help teachers develop mathematically proficient students (see Tool 3.7, which can be used with a video or actual observation).

    b. *Procedural knowledge is equivalent to drill and practice.* Distinguishing between procedural knowledge and drill is important. Drill might lead to weak procedural knowledge, while experiences selecting and using strategies and learning why algorithms work can develop strong procedural knowledge. Drill can provide opportunities for students to practice strategy selection and flexibility, as can discussions of the problems. The Overview addresses this misconception (see also Tool 3.3).

● Coaching Lessons From the Field

I was working with a group of emerging coaches related to implementing worthwhile tasks with their teachers. The coaches each selected what they felt was a worthwhile task. Next, they selected one of the planning tools related to worthwhile tasks (see Tools 3.1, 3.2, 3.5, and 3.6). Then, I explained that they were to imagine they were a coachee/teacher and they had received this tool from their coach and were asked to complete it before meeting for a planning conversation so that it could be used as a talking guide. This activity helped these future coaches think about the importance of providing "think time" for teachers.

—Robyn Ovrick
University of Georgia

Connecting to the Leading for Mathematical Proficiency (LMP) Framework

As teachers focus on content knowledge and worthwhile tasks, it is important for them to make explicit connections to the *Shifts in Classroom Practice* and the Mathematical Practices. The brief paragraphs that follow provide ideas for making these connections. Tool 3.1 can also be used for connecting to the *Shifts*.

● Connecting Content and Worthwhile Tasks to Shifts in Classroom Practice

| Shifts | 1 | 2 | 3 | 4 | 5 | 6 | 7 | 8 |

Shift 2: From *routine tasks* toward *reasoning tasks*

Teacher uses tasks involving recall of previously learned facts, rules, or definitions and provides students with specific strategies to follow. ———→ Teacher uses tasks that lend themselves to multiple representations, strategies, or pathways encouraging student explanation (how) and justification (why/when) of solution strategies.

Having a strong understanding of mathematics enables a teacher to "see" the mathematics in a task and consider the possible ways in which it could be represented and solved. Clearly, this impacts their abilities to select and to implement tasks effectively. Content knowledge and task selection go hand in hand: Considering the quality of a task can enhance teacher content knowledge, and studying content more deeply can similarly help teachers select better tasks (see Tools 3.2, 3.5, 3.6, 3.7, and 3.10).

Shift 3: From *teaching about representations* toward *teaching through representations*

Teacher shows students how to create a representation (e.g., a graph or picture). ———→ Teacher uses lesson goals to determine whether to highlight particular representations or to have students select a representation; in both cases, teacher provides opportunities for students to compare different representations and how they connect to key mathematical concepts.

Representations are not a learning goal in and of themselves; they serve to make mathematics more visible and to help students come to understand such things as why an algorithm works. Consider, for example, how partitioning a rectangle can help students visualize multiplication for whole numbers, decimals, fractions, and algebraic expressions. Here again, we see the importance of teachers understanding such representations and concepts themselves so that they can find tasks with potential and then navigate classroom discussions that connect representations to concepts and procedures (see Tool 3.4).

You can engage teachers in these *Shifts* by focusing specifically on content knowledge or on worthwhile tasks (or both together; see Tool 3.1). There are lots of possibilities for this Focus Zone! For example, with a focus on content knowledge, you might select *Shift* 3, ask teachers to consider the representations for an upcoming unit or lesson, and then consider what teaching moves look like across the continuum. Additionally, or instead, you could focus the discussion on fluency using both *Shifts 3* and *6*, selecting a procedural topic that is on the horizon, and ask teachers to think of things that they might say or do that fall across the continua. To focus on worthwhile tasks, there are many options. You might have them bring a copy of a student page they plan to use, cut up the page into tasks, and place each task on the continuum in terms of its potential to focus on reasoning or to support learning through representations. Or the tasks can be explored in terms of four considerations: what representations are possible, what strategies are possible, what high-level questions fit the task, and to what extent might students grapple with the problem (use a piece of paper folded into fourths to record ideas). When teachers struggle to answer these four questions for a task or worksheet, it can open up a rich conversation about the quality of that task or worksheet. Then, discussion can focus on how to adapt the task(s) or whether to replace them entirely.

Ideas in the shaded boxes provide a glimpse of exactly how to engage teachers with these important *Shifts*.

Connecting Content and Worthwhile Tasks to Mathematical Practices

MPs | 1 | 2 | 3 | 4 | 5 | 6 | 7 | 8

Key connections between the Mathematical Practices and Focus Zones add another dimension to the discussion.

It is easy to argue that in developing content knowledge, all the Mathematical Practices are relevant, and, in fact, this is certainly true. Here, we describe connections for just four and how they connect to content knowledge and worthwhile tasks.

1. ***Make sense of problems and persevere in solving them***. Students must have strong conceptual and procedural knowledge to be able to consider different solution pathways. Additionally, students must have frequent opportunities to engage in higher-level thinking related to all mathematical topics that they learn. As illustrated in Figure 3.2, this will not happen if they are simply told, "This is the procedure for this type of problem—do this."

This section provides a menu of discussion questions that might be used in professional learning or one-on-one coaching.

Coaching Questions for Discussion

Questions Related to the Focus Zone: Content Knowledge and Worthwhile Tasks

1. As you focus on a selected content standard, what might you do to connect it to other content the students have learned or will be learning?

2. Because we have likely learned from and taught from curricula that overemphasize procedures without connections, we need to be intentional about how we approach procedural fluency. What strategies might you use to determine the related concepts and procedures and ensure that you have a good balance of the two?

3. If you are focusing on developing procedural fluency, what might the student and teacher behaviors look like? What might the sample tasks look like? What might the student discourse sound like? How will you connect it to their conceptual knowledge?

4. What does it mean to have a task that is cognitively demanding? What features will it have or not have? What adaptations can you make to a task (in general) that can raise or lower the level of cognitive demand?

5. In considering the objectives of a particular lesson, in what ways do the lesson objectives include conceptual and procedural knowledge and making connections between them? What level of cognition (level of thinking) is expected?

This section points to additional resources you might want to use in a book study, lesson study, short course, or otherwise.

Where to Learn More

Books

NCTM Essential Understanding Series and *Putting Essential Understandings Into Practice* Series (www. nctm.org).

> *These are a series of short books that address a particular topic (e.g., ratios, proportions, and proportional reasoning), identifying the big ideas, connections of the idea to other mathematics, and instructional recommendations. As the titles imply, the first series uses examples and illustrations about the mathematics, and the second series focuses more on implementing the ideas in classrooms.*

National Council of Teachers of Mathematics. (2012). *Rich and Engaging Mathematical Tasks: Grades 5–9*. Reston, VA: NCTM.

> *This book contains a collection of articles and worthwhile mathematical tasks for teachers to use with their students to promote the understanding of the mathematical content highlighted in the Common Core State Standards for Mathematics.*

Schrock, C., Norris, K., Pugalee, D., Seitz R., & Hollingshead, F. (2013). *Great Tasks for Mathematics K–5 and 6–12: Engaging Activities for Effective Instruction and Assessment That Integrate Content and Practices of the CCSS-M*. Aurora, CO: NCSM.

> *The strong collection of tasks in these two books focus on important content and are designed to promote mathematical reasoning. The tasks have also been implemented in classrooms.*

Schuster, L., & Anderson, N.C. (2005). *Good Questions for Math Teaching: Why Ask Them and What to Ask (5–8)*. Sausalito, CA: Math Solutions.

Sullivan, P., & Lilburn, P. (2002). *Good Questions for Math Teaching: Why Ask Them and What to Ask (K–6)*. Sausalito, CA: Math Solutions.

> *While the title of these books says "good questions," it really refers to "good tasks"—and lots of them! You can find many activities that would be great opening tasks for a workshop setting, or particular tasks from the book could be identified for use in an upcoming lesson and could be the focus of a planning session.*

Articles

Drake, C., Land, T. J., Bartell, T. G., Aguirre, J. M., Foote, M. Q., McDuffie, A. R., & Turner, E. E. (2015). "Three Strategies for Opening Curriculum Spaces." *Teaching Children Mathematics, 21*(6), 346–353.

> *Though it uses elementary curriculum as an example, this article can be used at all levels. The authors share how to connect students' mathematical knowledge bases and increase the meaning-making in a lesson by rearranging lesson components, adapting tasks, and making authentic connections.*

Lange, K. E., Booth, J. L., & Newton, K. J. (2014). "Learning Algebra From Worked Examples." *Mathematics Teacher, 107*(7), 534–540.

McGinn, K. M., Lange, K. E., & Booth, J. L. (2015). "A Worked Example for Creating Worked Examples." *Mathematics Teaching in the Middle School, 21*, 26–33.

Coach's Toolkits in each
chapter support teacher
learning with self-assessment,
planning, data gathering, and
reflecting tools.

Coach's Toolkit

These tools are a menu from which you can select any that make sense for your setting/context. They can be
used independently or as part of a coaching cycle. You may start with the self-assessment, which can guide
you in deciding which of the other tools may be most useful. If using these tools for a coaching cycle, mix and
match as you like, or use one of the combinations we suggest in the diagrams that follow. The tools include
instructions to the coach and the teacher.

Self-Assess

3.1	Connecting *Shifts* to Content and Worthwhile Tasks Self-Assessment

Gather Data

3.7	Developing Mathematical Proficiency
3.8	Implementing Cognitively Demanding Tasks

Plan

3.2	Connecting *Shifts* to Content and Tasks
3.3	Focus on Fluency
3.4	P.I.C.S. Page
3.5	Analyzing Level of Cognitive Demand
3.6	Worthwhile Task Analysis

Reflect

3.9	Impact on Students' Emerging Fluency
3.10	Reflecting on Task Implementation

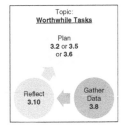

Topic:
Worthwhile Tasks

Plan
3.2 or **3.5**
or **3.6**

Reflect
3.10

Gather
Data
3.8

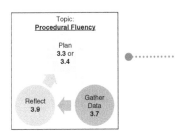

Topic:
Procedural Fluency

Plan
3.3 or
3.4

Reflect
3.9

Gather
Data
3.7

Additional Tools in Other Chapters

Implementing high-quality tasks is also the focus of Chapter 5 (Questioning and Discourse; see Tools 5.3, 5.4,
and 5.5), Chapter 7 (Analyzing Student Thinking), and Chapter 8 (Differentiating Instruction, which includes
differentiating tasks).

online
resources | To download the coaching tools for Chapter 3 that only have instructions
for the teacher, go to **resources.corwin.com/mathematicscoaching**.

Over 90 tools are
downloadable for
free from the book's
companion website.

You'll find
suggestions for how
to use specific tools
in a coaching cycle,
depending on your
goals.

5.1 Connecting *Shifts* to Questioning and Discourse
Self-Assessment

Instructions to Coach: Ask teachers (individually or as part of a PLC activity) to self-assess where they position themselves on each of these Shifts in Classroom Practice related to questioning and discourse. Use the questions that follow during a coaching conversation or in a PD setting to support teachers (and to help you decide which tools may be most useful from this chapter). A one-page version of this tool is available for download.

Instructions: The following *Shifts in Classroom Practice* have specific connections to questioning and discourse. Put an *X* on the continuum of each *Shift* to identify where you currently see your practice.

Tool 5.1 Shifts

Shift 4: From *show-and-tell* toward *share-and-compare*

| Teacher has students share their answers. | → | Teacher creates a dynamic forum where students share, listen, honor, and critique each other's ideas to clarify and deepen mathematical understandings and language; teacher strategically invites participation in ways that facilitate mathematical connections. |

Shift 5: From *questions that seek expected answers* toward *questions that illuminate and deepen student understanding*

| Teacher poses closed and/or low-level questions, confirms correctness of responses, and provides little or no opportunity for students to explain their thinking. | → | Teacher poses questions that advance student thinking, deepen students' understanding, make the mathematics more visible, provide insights into student reasoning, and promote meaningful reflection. |

Shift 7: From *mathematics-made-easy* toward *mathematics-takes-time*

| Teacher presents mathematics in small chunks so that students reach solutions quickly. | → | Teacher questions, encourages, provides time, and explicitly states the value of grappling with mathematical tasks, making multiple attempts, and learning from mistakes. |

Shift 8: From *looking at correct answers* toward *looking for students' thinking*

| Teacher attends to whether an answer or procedure is (or is not) correct. | → | Teacher identifies specific strategies or representations that are important to notice; strategically uses observations, student responses to questions, and written work to determine what students understand; and uses these data to inform in-the-moment discourse and future lessons. |

Tool 5.1 Reflection Questions

1. What do you notice, in general, about your self-assessment of these *Shifts in Classroom Practice*?
2. What might be specific teaching moves that align with where you placed yourself on the *Shifts*?
3. What might be specific teaching moves that align *to the right of* where you placed yourself on the *Shifts*?
4. What might be some professional learning opportunities to help you move to the right for one or more of these *Shifts*?

Self-Assessments help teachers pinpoint and reflect on where they are on the *Shifts in Classroom Practice* continuum or each focus zone.

 3.2 Connecting *Shifts* to Content and Tasks

Instructions to the Coach: As an alternative or as a follow-up to Tool 3.1, use this tool to help teachers see how the Shifts apply to a particular lesson. In a coaching cycle, teachers could complete this in preparation for a planning conversation, or they could just bring their task with them and the planning conversation can be the questions provided in the template.

Instructions: The *Shifts in Classroom Practice* can provide support in keeping the level of cognitive demand high through task implementation. Select one or two *Shifts* as a focus and then complete the planning table here.

Shift(s):

Toward ... _____

Toward ... _____

The Task	Classroom Environment	Setting Up the Task
How might you adapt the task to reflect the selected *Shifts*?	How might you organize students?	How might you pose the task to reflect the selected *Shift(s)*?

Lesson	Discussing the Task	Assessing
How might you structure the lesson itself to align with the *Shift(s)*?	What questions or questioning strategies will you use?	What will you be looking for as students work?

Source: *Previously published by Bay-Williams, J., McGatha, M., Kobett, B., and Wray, J. (2014). Mathematics Coaching: Resources and Tools for Coaches and Leaders, K–12. New York, NY: Pearson Education, Inc.*

Planning tools prompt teachers to brainstorm and problem-solve coaching challenges.

Data gathering tools help teachers capture key measures of coaching progress and classroom instruction to inform practice.

 3.6 Worthwhile Task Analysis

Instructions to the Coach: This tool is very effective in professional learning to focus on analyzing and adapting tasks. And it can be used in a coaching cycle.

Instructions: Using the Worthwhile Tasks Focus Zone prompts below (NCTM, 1991; NCTM, 2007), rate a task you are planning to use in a lesson. Add comments about how it might be adapted to better address the stated quality of a worthwhile task.

1 = No evidence of the quality in the task, or it is not possible to address this quality with this task.

2 = The quality is evident in minor ways, or incorporating it is possible.

3 = The quality is evident in the task.

4 = The quality is central to the task and is important to the success of the lesson.

Aspects of a Worthwhile Task	Rating				How I Might Enhance Task
Mathematics in the task is powerful.					
1. Is grade or course-level appropriate	1	2	3	4	
2. Makes connections between concepts and procedures (high cognitive level)	1	2	3	4	
3. Makes connections between different mathematical topics	1	2	3	4	
4. Requires reasoning (non-algorithmic thinking)	1	2	3	4	
Task is connected to the student.					
5. Connects to real situations that are familiar and relevant to them	1	2	3	4	
6. Provides multiple entry points that make it accessible to each student	1	2	3	4	
7. Is appropriately challenging (engages students' interests and intellect)	1	2	3	4	
Task lends to observing and assessing student understanding.					
8. Provides multiple ways to demonstrate understanding of the mathematics	1	2	3	4	
9. Requires students to illustrate or explain mathematical ideas	1	2	3	4	
10. Has potential to develop perseverance and positive student dispositions	1	2	3	4	

1. Describe your overall evaluation of whether this task/lesson has the potential to engage students in higher-level thinking.

2. What adaptations can you make to the task or lesson to increase its higher-level thinking potential?

Reflecting tools engage teachers in considering the impact of specific instructional moves and related implications for practice.

 3.10 Reflecting on Task Implementation

Instructions to the Coach: This tool can be used as a follow-up to Tool 3.8. If not connected with Tool 3.8, begin by having the teacher review and sort teacher actions and student actions from the lesson.

Instructions: Discuss the following, using evidence and examples from the data collected.

1. In what ways did you feel that the implementation of the task resulted in its being a "worthwhile" opportunity for students to learn important mathematics?

2. How effective was the implementation of the task in engaging students with the context of the problem?

3. How effective was the implementation of the task in helping students understand concepts, procedures, and connections between the two?

4. To what extent was each student appropriately challenged and engaged in productive struggle?

5. In looking at the data, what teacher moves did you make that may have maintained a high level of cognitive demand and/or reduced the high level of cognitive demand? What was effective?

6. What general insights about Effective Mathematics Teaching did this lesson cycle provide?

Source: *Previously published by Bay-Williams, J., McGatha, M., Kobett, B., and Wray, J. (2014). Mathematics Coaching: Resources and Tools for Coaches and Leaders, K–12. New York, NY: Pearson Education, Inc.*

Everything You Need for
MATHEMATICS
COACHING

GRADES K–12

Everything You Need for
MATHEMATICS
COACHING

Tools, Plans, and a Process That
Works for Any Instructional Leader

90+ Tools
Inside & Online!

Maggie B. McGatha + Jennifer M. Bay-Williams
with Beth McCord Kobett + Jonathan A. Wray

A JOINT PUBLICATION

NATIONAL COUNCIL OF
TEACHERS OF MATHEMATICS

FOR INFORMATION:

Corwin
A SAGE Company
2455 Teller Road
Thousand Oaks, California 91320
(800) 233-9936
www.corwin.com

SAGE Publications Ltd.
1 Oliver's Yard
55 City Road
London, EC1Y 1SP
United Kingdom

SAGE Publications India Pvt. Ltd.
B 1/I 1 Mohan Cooperative Industrial Area
Mathura Road, New Delhi 110 044
India

SAGE Publications Asia-Pacific Pte. Ltd.
3 Church Street
#10-04 Samsung Hub
Singapore 049483

Printed in Canada.

Program Manager, Mathematics: Erin Null
Editorial Development Manager: Julie Nemer
Interim Associate Editor: Cynthia Gomez
Editorial Assistant: Jessica Vidal
Production Editor: Tori Mirsadjadi
Copy Editor: Amy Hanquist Harris
Typesetter: Integra
Proofreader: Vicki Rosenzweig
Indexer: May Hasso
Cover and Interior Designer: Gail Buschman
Marketing Manager: Margaret O'Connor

ISBN: 978-1-5443-1698-7

This book is printed on acid-free paper.

MIX
Paper from responsible sources
FSC® C103567

18 19 20 21 22 10 9 8 7 6 5 4 3 2 1

Contents

PART I — Creating a Road Map: Big Ideas of Coaching, Teaching, and Learning

PART II — Exploring Zones on the Journey: Professional Learning Focus Areas

Chapter 4
ENGAGING STUDENTS 71

Chapter 5
QUESTIONING AND DISCOURSE 93

Part III   Navigating a Successful Journey: Strategies and Tools for You, the Coach

Chapter 12
PRESENTING PROFESSIONAL DEVELOPMENT   249

Chapter 13
FACILITATING PROFESSIONAL LEARNING COMMUNITIES   277

Appendix:
Bookmarks   301

Preface

We wrote this book for the very busy mathematics coach (by the term *coach*, we include all who provide leadership to others who are teaching mathematics in K–12 classrooms). We imagine that a busy coach may not have time to read a preface! But we want to introduce you, as briefly as we can, to what this book has in store for you and why we wrote it in the first place. So let's start there.

Why did we write this book?

We think—in fact, we know—that mathematics coaches make a positive difference in students' mathematical learning experiences. We also know that coaches often function in relative isolation, having to invent, create, react, and respond in myriad situations, often with little time to prepare. There are many resources out there, and it is tough to know which ones to use and for which purposes. Meanwhile, you might be providing professional development, facilitating professional learning communities (PLCs), or leading a coaching cycle and wondering many things: What format might you use? How might you identify a focus for that professional development or lesson cycle? Once a topic is chosen, what aspects of that topic might be most impactful? This dizzying list only scrapes the surface of what crossroads you might come to in your journey toward improving mathematics teaching and learning in your setting.

What do we have in store for you?

We hope this is your go-to resource to navigate your journey as a mathematics coach! We have surveyed many coaches and leaders to identify must-know topics, skills, and issues and then used those insights in selecting topics for this book. Think of it as a compendium of mathematics coaching *Reader's Digests*! In other words, each chapter is a condensed explanation of critical mathematics coaching actions and critical teaching actions along with tools, tools, and more tools to help you do the work: tools for navigating a coaching conversation; tools for planning, facilitating, and evaluating professional development; and so on. Finally, the back of the book includes an Appendix with two bookmarks for your use and a glossary of key words used throughout the chapters (many education words/phrases take on multiple meanings, so we wanted to make sure you knew how we use terms in this book). Words that appear in the glossary are bolded the first time they are discussed in the text. Here, we share the book's features in a little more detail.

Part I: Creating a Road Map: Big Ideas of Coaching, Teaching, and Learning

What is the purpose of your coaching? We hope your answer is something like "helping teachers to develop mathematically proficient students." Learning goals for students *must* be the clear purpose of all things coaching. So we begin with our Leading for Mathematical Proficiency (LMP) Framework, which positions student outcomes as the focus of your work as the coach, as well as teachers' professional learning goals. The Framework explicitly connects Mathematical Practices for students with Effective Mathematics Teaching Practices for teachers.

Important to Part 1 of the book is our *Shifts in Classroom Practice.* If you know our first book, you know that these *Shifts* are a guide for teachers as individuals or as part of a PLC, reflecting on where they are on the continuum and then working to move toward more effective practices. Our *Shifts in Classroom Practice* in this book are now a one-to-one match to the NCTM Effective Mathematics Teaching Practices (NCTM, 2014). All of this is explained more clearly in Chapter 1.

Discussions and tools in Chapter 2 focus on the big ideas of the Framework—helping teachers think about the Mathematical Practices and Effective Mathematics Teaching Practices. These tools—all available online—are flexible. You may like one for a particular teacher and not want to use it with another. These tools may also be used in connection with any of the Part II chapters.

 To download the coaching tools for Chapter 2, go to **resources.corwin.com/ mathematicscoaching**.

Part II: Exploring Zones on the Journey: Professional Learning Focus Areas

Once you have engaged teachers with the big goals, it is time to start exploring an identified focus area, or focus zone. Focus Zones include topics such as engaging students, using formative assessment, and differentiating instruction. If you have our first book, *Mathematics Coaching: Resources and Tools for Coaches and Leaders, K–12,* you will see that these focus areas have not changed much (we added a new chapter on engaging students), but the format of each chapter has changed quite a bit, and the tools are new or improved. Each of these chapters (3–10) has an identical format, shown here:

In the Coach's Digest . . .

 To download the Overview for each chapter to share with teachers, go to **resources.corwin.com/mathematicscoaching**.

Overview. This section is unique in that the audience is you *and* your teachers. You can read it for your own learning and can also download a version of it for sharing with your teachers. Remember it is a *Reader's Digest*-style coverage, so you may want to supplement or simply use it as a springboard to hear from teachers what else they know about this Zone!

Coaching Considerations for Professional Learning. Here, we offer ideas for how you might engage teachers in learning about this particular Focus Zone. Some of the ideas refer to professional development sessions, while others refer to coaching cycles. We tried to provide a "menu" from which you might find something that sounds good.

Coaching Lessons From the Field. We have heard many stories from many coaches during the five years that our first book has been "out there." This book has similar focus areas, so these stories provide you with some ideas about how coaches are supporting teachers in their settings.

Connecting to the Framework. It is worth saying again: The journey to mathematics proficiency must include consistent and frequent connections to the big goals of effective teaching (our *Shifts*) and student outcomes (Mathematical Practices). We share important connections between the *Shifts* and the Focus Zones and the Mathematical Practices and Focus Zones, along with some ideas for helping you engage teachers with these important connections.

Coaching Questions for Discussion. Just as it sounds, this section provides a menu of discussion questions. They might be used in professional learning or one-on-one coaching. And the final set of questions connects to the *Shifts* and the Mathematical Practices.

Where to Learn More. This section might provide sources for more reading for you or might point you toward readings that you want to use in a book study, lesson study, short course, or otherwise. Brief descriptions are offered—we hope they do justice to the fantastic articles, books, and websites on these Focus Zones!

To download the coaching tools for each chapter in Part II, go to **resources.corwin.com/mathematicscoaching**.

In the Coach's Toolkit . . .

Each chapter has a set of tools to support teacher learning. The first tool in every chapter is a self-assessment tool, and it includes selected *Shifts in Classroom Practice* (in contrast to the self-assessment in Chapter 2 that focuses on all the *Shifts*). We designed the remaining tools to help plan, gather data, and reflect. If you are engaged in a coaching cycle, then you can pick one from each category. Guidance for which ones you might put together is provided on the cover page. Also, a tool in one chapter might be a great fit for efforts in another Focus Zone. Use it! Adapt it! We offer some suggestions of ones we don't want you to miss, but there are many tools, and that is so that you can find ones that you think will work best for you and your teachers.

A final note: These Focus Zones are in no particular order. They are just zones on the journey. As you will see from the *Shifts* and Mathematical Practices addressed in each Focus Zone, they all have the potential to increase effective teaching practice and student mathematical proficiency.

Part III: Navigating a Successful Journey: Strategies and Tools for You, the Coach

How does one avoid or minimize road bumps along the way? This is hard to do when the journey may be in uncharted territory, but this part of the book is here to assist!

The first chapter (Chapter 11) offers you guidance on communicating with colleagues. If you have not established rapport or good listening skills, for example, your journey will be long and difficult. Chapters 12 and 13 are designed to help you *lead* the way (i.e., lead a professional development session) and *guide* teachers along the way (i.e., facilitate a professional learning community). More tools? You got it!

Acknowledgments

We offer our sincere gratitude to the many people in our professional lives who have helped us grow as mathematics teachers and leaders. We have each been influenced by so many teachers and mentors, and we continue to learn so much from our ongoing work with teachers and coaches.

We want to say a special thank you to those of you who submitted stories about ways you have used our mathematics coaching tools: Tracy Beck, Sandra Davis, Natasha Crissey, Linda Fulmore, Melisa Hancock, Rita Hays, Keith Kerschen, Maria Leaman, Robyn Orvick, Denise Porter, Ryann N. Shelton, Jeanne Simpson, Candy Thomas, and Trena Wilkerson.

Additionally, our thanks to each of the following mathematics instructional support teachers and mathematics resource teachers in Howard County, Maryland, for their valuable input on the tools in this book.

Meredith Adams	Lindsay Kelley	Shanti Stone
Holly Cheung	Andrea Lang	AnnMarie Varlotta
Matthew Cox	Anne Metzbower	Melissa Waggoner
Christiana Dellota	Jennifer Mullinix	Robin White
Jenna Demario	Damitra Newsome	Tara Wilson
Megan Gittermann	Greta Richard	Elizabeth Zinger

Maggie B. McGatha works full-time with coaches, teacher leaders, and administrators as a training associate for Cognitive Coaching℠ and Adaptive Schools. At the University of Louisville, she teaches courses on coaching and supporting collaborative groups, along with advanced-mathematics methods courses. Maggie is a former middle school mathematics teacher, the author of numerous articles, and a coauthor of *Mathematics Coaching: Resources and Tools for Coaches and Leaders, K–12* and *On the Money: Math Activities to Build Financial Literacy, Grades 6–8.* She is active in state and national mathematics organizations, including currently serving as the vice president for the membership division (2017–2020) of the Association of Mathematics Teacher Educators (AMTE) and as a member of the advisory board for the Mathematics Institute of Wisconsin. She received her doctorate from Vanderbilt University (Tennessee). Maggie enjoys traveling and spending time with her family, especially her grandkids!

Jennifer M. Bay-Williams is a professor at the University of Louisville (Kentucky). She is a leader in mathematics education, committed to supporting teachers and teacher leaders in their efforts to improve mathematics learning opportunities for every student. She has authored many articles in NCTM journals and coauthored many books, including *Teaching Elementary and Middle School Mathematics: Teaching Developmentally,* the related three-book series, *Teaching Student-Centered Mathematics* (with John Van de Walle, LouAnn Lovin, and Karen Karp), *On the Money* (financial literacy book series), *Developing Essential Understanding of Addition and Subtraction in Prekindergarten–Grade 2*, and *Math and Literature: Grades 6–8.* Jennifer is very active in national mathematics organizations, having served on the NCTM Board of Directors, as the Association of Mathematics Teacher Educators (AMTE) secretary and president, and as lead writer for the *Standards for the Preparation of Teachers of Mathematics.* She currently serves on the board of directors for TODOS: Mathematics for ALL. She received her PhD from the University of Missouri (Columbia), taught K–12 mathematics in Missouri and Peru, and continues to work in schools supporting effective mathematics teaching. She enjoys the beach, watching her daughter play softball, playing tennis with her son, and seeing pretty much any Broadway production with her husband or whoever wants to go!

Beth McCord Kobett is an associate professor in the School of Education at Stevenson University, where she works with preservice teachers and leads professional learning efforts in mathematics education both regionally and nationally. She is also the lead consultant for the Elementary Mathematics Specialists and Teacher Leaders Project. She is a former classroom teacher, elementary mathematics specialist, adjunct professor, and university supervisor. At the undergraduate level, Beth teaches early childhood, elementary, and middle school mathematics methods courses, as well as equity in education courses. At the graduate level, she teaches courses on diagnostics and interventions for mathematics students. She has a particular interest in developing problem-solving dispositions in students through problem-based teaching. Dr. Kobett is a recipient of the Mathematics Educator of the Year Award from the Maryland Council of Teachers of Mathematics (MCTM). She also received Stevenson University's Excellence in Teaching Award as both an adjunct and full-time faculty member. Beth received her master's and doctoral degrees from Johns Hopkins University and bachelor's degree from the University of Missouri. She enjoys running and spending time with her family.

 **Jonathan A. Wray** is the secondary mathematics curriculum supervisor for the Howard County Public School System (Maryland). He is also the project manager of the Elementary Mathematics Specialists and Teacher Leaders (ems&tl) Project. Jon has served as an elected member of the National Council of Teachers of Mathematics (NCTM) Board of Directors. He is a former classroom teacher, mathematics resource teacher, and past president of both the Association of Maryland Mathematics Teacher Educators (AMMTE) and the Maryland Council of Teachers of Mathematics (MCTM). Mr. Wray is a recipient of the MCTM Outstanding Teacher Mentor Award and was selected as his district's Outstanding Technology Leader in Education by the Maryland Society for Educational Technology. His interests include the leadership roles of mathematics coaches/specialists; teacher collaboration; effective and engaging teaching and learning strategies; strategic uses of technology; and issues related to access, equity, and student empowerment in mathematics education. Jon received his master's degree from Johns Hopkins University and bachelor's degree from Towson University. Jon loves being wherever his family is, watching his children grow, and making fun memories together.

Part I

Creating a Road Map

Big Ideas of Coaching, Teaching, and Learning

There is a lot to know and do as a mathematics coach! Having a clear purpose, a cohesive plan, and important outcomes is essential, so we believe that *Everything You Need for Mathematics Coaching* must begin with a leadership framework—one that connects important student learning outcomes (mathematical practices) to effective teaching practices (*Shifts in Classroom Practice* and NCTM Effective Mathematics Teaching Practices). This Leading for Mathematical Proficiency (LMP) Framework can guide your work. Also, consider what you might need to help teachers in their journey to implement effective teaching practices that support the development of mathematical practices: plans, discussions, and tools! Chapter 2 is focused on this very topic. These first two chapters truly are the place to begin—not just with reading this book, but with designing a focused professional learning plan for you and your teachers. We hope this road map, along with knowledge about the coaching cycle, will empower you to lead your teachers with clarity and purpose!

Chapter 1

Coaching for Mathematical Proficiency

Ultimately, the goal of every mathematics coach is to improve the chances that students become *competent* and *confident* in mathematics. This simply stated purpose is anything but simple to accomplish. Each coach or leader works within the context of his or her own setting to decide how to best focus efforts toward this goal. In one setting, a coach might focus on helping teachers to learn content more deeply, considering the way in which mathematical ideas are connected within their grade or course and across grades or courses. In another setting, the focus may be on teaching, with efforts to consider how particular instructional moves provide or prevent opportunities for each student to engage with the content. If you have been coaching for a while, you might feel that your work is anything but focused, with various initiatives happening in different schools or classrooms. But everything that you are doing is in some way or another focused on **mathematical proficiency** for every student. The purpose of this book is to help you, the leader (and to help you help your teachers), to see the connections between your **professional learning** activities, teaching, and student learning. *Everything You Need for Mathematics Coaching: Tools, Plans, and a Process That Works for Any Instructional Leader* might be thought of as your *Reader's Digest*, providing you with a shortened version of critical topics in coaching and teaching (e.g., facilitating discourse) and a collection of resources to assist you in working on that topic.

Recognizing the busy lives of coaches, we try to boil down the discussion of these topics to the bare necessities, offer you a menu of places to go for more information, and provide a collection of tools that can be used for professional learning or for one-on-one coaching (the Preface elaborates on each of these features, and each chapter offers an opening note to guide you to what you need). In the chapters that follow, we zoom in on specific topics; in this first chapter, that abbreviated discussion is about the big picture. The second chapter also focuses on the big picture, with an overview addressed to teachers (that you can download and share), along with tools to support their learning. The remainder of Chapter 2 is addressed to you, the coach or instructional leader.

Leading for Mathematical Proficiency (LMP) Framework

It is natural, even wise, to seek justification and rationale when asked to change. When thinking of a change in teaching practice, there are many possible ways to justify it. Learning of research that something worked, hearing about a situation in which a practice "worked," or hearing a testimonial about an idea that really engaged students are ways that some teaching practices get picked up and tried. We would like to suggest that fundamentally we must start at what it is we want students to be able to do—not which standards they will learn, but what we want them to do if we are truly preparing them to do mathematics. What it means to do mathematics is best described in the Mathematical Practices (NGA & CCSSO, 2010).

The Leading for Mathematical Proficiency (LMP) Framework (see Figure 1.1) frames the goals of professional learning on developing mathematical proficiency in all students. The LMP Framework is dually focused to both help you, as coaches, see how pieces of your efforts fit together in a purposeful manner and to help you, as facilitators, communicate these connections to teachers and other stakeholders. The bottom line is that we must regularly revisit these connections to ensure that there is a clear purpose and cohesion to the activities that are occurring.

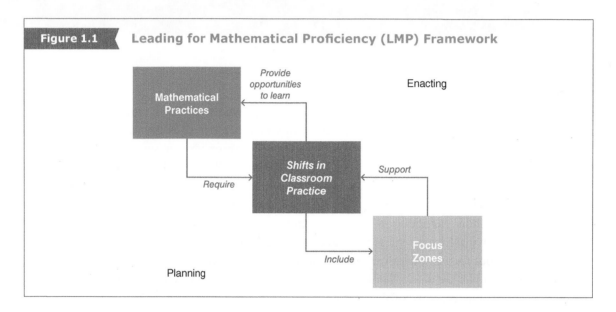

Figure 1.1 Leading for Mathematical Proficiency (LMP) Framework

Each of the major components of the LMP Framework has lists of more specific targets. Specifically, there are eight Mathematical Practices and eight *Shifts in Classroom Practice*. A coach may choose to focus on any number of focus zones, but we have identified eight. These are related to each other in complex ways and are presented in Figure 1.2 as an advanced organizer of the discussion of these major components.

Mathematical Practices

From the Common Core State Standards for Mathematics (NGA & CCSSO, 2010), and now included in many state standards, these expectations for students must be the ultimate goal of **mathematics coaching. Mathematical practices** describe what **mathematically proficient students** are able to do, and those descriptions will endure even if the labels are changed. Figure 1.3 lists each Mathematical Practice, along with excerpts from the descriptions (only the sentences beginning with "Mathematically proficient students ..." are included).

To place the Mathematical Practices as the goal of professional learning, as in the LMP Framework, professional learning must begin by asking,

What does each of the Mathematical Practices look like in action?

As New York Yankee player and coach Yogi Berra put it, "If you don't know where you are going, you'll end up someplace else." This advice for baseball also applies to instructional coaching! Related to the Mathematical Practices, you have to be sure that there is a shared understanding of what a student is doing if the student is modeling with mathematics (Mathematical Practice 4) or looking for structure

---

Figure 1.2 At-a-Glance Elements Within Each Component of the LMP Framework

Mathematical Practices

1. Make sense of problems and persevere in solving them.
2. Reason abstractly and quantitatively.
3. Construct viable arguments and critique the reasoning of others.
4. Model with mathematics.
5. Use appropriate tools strategically.
6. Attend to precision.
7. Look for and make use of structure.
8. Look for and express regularity in repeated reasoning.

Shifts in Classroom Practice

...Toward...

1. Communicates learning expectations
2. Reasoning tasks
3. Teaching through representations
4. Share-and-compare
5. Questions illuminate and deepen student understanding
6. Selecting efficient strategies
7. Mathematics-takes-time
8. Looking for students' thinking

Focus Zones

1. Content Knowledge and Worthwhile Tasks
2. Engaging Students
3. Questioning and Discourse
4. Formative Assessment
5. Analyzing Student Work
6. Differentiating Instruction for All Students
7. Supporting Emergent Multilingual Students
8. Supporting Students with Special Needs

(Mathematical Practice 7). The Mathematical Practices provide specific descriptors or "**look fors**" related to student actions, and these can and should be tied to the content that students are learning. In the Appendix, you will find the Mathematical Practices & Student Look Fors Bookmark, providing an at-a-glance resource for **professional development, lesson cycles**, and personal reference.

 You can also download the bookmark at **resources.corwin.com/mathematicscoaching.**

Even if the list of eight Mathematical Practices is familiar, being able to think about what they look like for a first grader or a tenth grader is not obvious. If teachers within a group have different interpretations of modeling, for example, then they are working toward different goals—which can make the work of PLCs confusing or ineffective. And if a teacher views these practices differently than the coach, it can interfere with the effectiveness of a lesson cycle. Chapter 2 provides numerous tools for helping to build a shared understanding of what the Mathematical Practices look like in action, and Chapter 12 provides several professional learning activities to engage groups of teachers in developing a shared understanding of these practices.

Shifts in Classroom Practice

With a vision of what a Mathematical Practice (or several Practices) looks like in terms of student actions or behaviors, the question becomes this:

How can we (as teachers and coaches) provide optimal learning opportunities for students to become mathematically proficient?

Creating opportunities for students to become mathematically proficient (i.e., demonstrate the Mathematical Practices) requires implementing teaching practices that focus on helping students acquire these practices while simultaneously working on creating deep understanding of the content. *Principles to Actions: Ensuring Mathematical Success for All* (PtA; NCTM, 2014) delineates eight **Effective Teaching Practices** (see Figure 1.3 on the next page). These practices are grounded in

| **Figure 1.3** | Mathematical Practices: What Mathematically Proficient Students Are Able to Do (NGA & CCSSO, 2010) |

1. Make sense of problems and persevere in solving them. Mathematically proficient students start by explaining to themselves the meaning of a problem and looking for entry points to its solution.... Mathematically proficient students can explain correspondences between equations, verbal descriptions, tables, and graphs or draw diagrams of important features and relationships, graph data, and search for regularity or trends.... Mathematically proficient students check their answers to problems using a different method, and they continually ask themselves, "Does this make sense?"

2. Reason abstractly and quantitatively. Mathematically proficient students make sense of quantities and their relationships in problem situations.

3. Construct viable arguments and critique the reasoning of others. Mathematically proficient students understand and use stated assumptions, definitions, and previously established results in constructing arguments.... Mathematically proficient students are also able to compare the effectiveness of two plausible arguments, distinguish correct logic or reasoning from that which is flawed, and—if there is a flaw in an argument—explain what it is.

4. Model with mathematics. Mathematically proficient students can apply the mathematics they know to solve problems arising in everyday life, society, and the workplace.... Mathematically proficient students who can apply what they know are comfortable making assumptions and approximations to simplify a complicated situation, realizing that these may need revision later.

5. Use appropriate tools strategically. Mathematically proficient students consider the available tools when solving a mathematical problem.... Mathematically proficient students are sufficiently familiar with tools appropriate for their grade or course and make sound decisions about when each of these tools might be helpful, recognizing both the insight to be gained and their limitations.

6. Attend to precision. Mathematically proficient students try to communicate precisely to others.

7. Look for and make use of structure. Mathematically proficient students look closely to discern a pattern or structure.

8. Look for and express regularity in repeated reasoning. Mathematically proficient students notice if calculations are repeated and look both for general methods and for shortcuts.

research, which is summarized not only in that important book but also in related reviews of research (Spangler & Wanko, 2017; see *Books* in the Where to Learn More section, later in this chapter).

For each of these Teaching Practices, we developed a *Shift in Classroom Practice*. A *Shift* succinctly describes the Teaching Practice along a continuum (see Figure 1.4). Teaching is a learning endeavor. There is always some way we can adapt our practice in order to better meet the needs of students. The complexity of teaching means novices are on a journey toward being more effective. Ongoing research in teaching and related fields such as brain research means that experienced teachers are also on a journey toward being more effective. In other words, we are all on the continuum somewhere and trying to move in the "right" direction. As a coach, you support teachers in self-assessing where their strengths lie and where they might want to shift their practices.

Figure 1.4   *Shifts in Classroom Practice*

Shift 1: From *stating-a-standard* toward *communicating expectations for learning*

Teacher shares broad performance goals and/or those provided in standards or curriculum documents.	→ Teacher creates lesson-specific learning goals and communicates these goals at critical times within the lesson to ensure students understand the lesson's purpose and what is expected of them.

NCTM Teaching Practice: **Establish mathematics goals to focus learning.** Effective teaching of mathematics establishes clear goals for the mathematics that students are learning, situates goals within learning progressions, and uses the goals to guide instructional decisions.

Shift 2: From *routine tasks* toward *reasoning tasks*

Teacher uses tasks involving recall of previously learned facts, rules, or definitions and provides students with specific strategies to follow.	→ Teacher uses tasks that lend themselves to multiple representations, strategies, or pathways encouraging student explanation (how) and justification (why/when) of solution strategies.

NCTM Teaching Practice: **Implement tasks that promote reasoning and problem-solving.** Effective teaching of mathematics engages students in solving and discussing tasks that promote mathematical reasoning and problem-solving and allow multiple entry points and varied **solution strategies**.

Shift 3: From *teaching about representations* toward *teaching through representations*

Teacher shows students how to create a representation (e.g., a graph or picture).	→ Teacher uses lesson goals to determine whether to highlight particular representations or to have students select a representation; in both cases, teacher provides opportunities for students to compare different representations and how they connect to key mathematical concepts.

NCTM Teaching Practice: **Use and connect mathematical representations.** Effective teaching of mathematics engages students in making connections among mathematical representations to deepen understanding of mathematics concepts and procedures and as tools for problem-solving.

Shift 4: From *show-and-tell* toward *share-and-compare*

Teacher has students share their answers.	→ Teacher creates a dynamic forum where students share, listen, honor, and critique each other's ideas to clarify and deepen mathematical understandings and language; teacher strategically invites participation in ways that facilitate mathematical connections.

NCTM Teaching Practice: **Facilitate meaningful mathematical discourse.** Effective teaching of mathematics facilitates discourse among students to build shared understanding of mathematical ideas by analyzing and comparing student approaches and arguments.

(Continued)

(*Continued*)

Shift 5: From *questions that seek expected answers* toward *questions that illuminate and deepen student understanding*

Teacher poses closed and/or low-level questions, confirms correctness of responses, and provides little or no opportunity for students to explain their thinking.	→	Teacher poses questions that advance student thinking, deepen students' understanding, make the mathematics more visible, provide insights into student reasoning, and promote meaningful reflection.

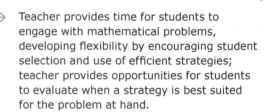

NCTM Teaching Practice: **Pose purposeful questions.** Effective teaching of mathematics uses purposeful questions to assess and advance students' reasoning and sense making about important mathematical ideas and relationships.

Shift 6: From *teaching so that students replicate procedures* toward *teaching so that students select efficient strategies*

Teacher approaches facts and procedures with the goal of speed and accuracy.	→	Teacher provides time for students to engage with mathematical problems, developing flexibility by encouraging student selection and use of efficient strategies; teacher provides opportunities for students to evaluate when a strategy is best suited for the problem at hand.

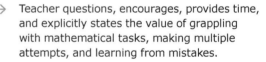

NCTM Teaching Practice: **Build procedural fluency from conceptual understanding.** Effective teaching of mathematics builds fluency with **procedures** on a foundation of **conceptual understanding** so that students, over time, become skillful in using procedures flexibly as they solve contextual and mathematical problems.

Shift 7: From *mathematics-made-easy* toward *mathematics-takes-time*

Teacher presents mathematics in small chunks so that students reach solutions quickly.	→	Teacher questions, encourages, provides time, and explicitly states the value of grappling with mathematical tasks, making multiple attempts, and learning from mistakes.

NCTM Teaching Practice: **Support productive struggle in learning mathematics.** Effective teaching of mathematics consistently provides students, individually and collectively, with opportunities and supports to engage in productive struggle as they grapple with mathematical ideas and relationships.

Shift 8: From *looking at correct answers* toward *looking for students' thinking*

Teacher attends to whether an answer or procedure is (or is not) correct.	→	Teacher identifies specific strategies or representations that are important to notice; strategically uses observations, student responses to questions, and written work to determine what students understand; and uses these data to inform in-the-moment discourse and future lessons.

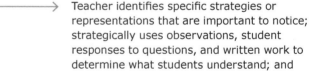

NCTM Teaching Practice: **Elicit and use evidence of student thinking.** Effective teaching of mathematics uses evidence of student thinking to assess progress toward mathematical understanding and to adjust instruction continually in ways that support and extend learning.

Source: NCTM Teaching Practice Statements from National Council of Teachers of Mathematics (NCTM). (2014). Principles to Actions: Ensuring Mathematical Success for All. *Reston, VA: Author.*

Importantly, the *Shifts in Classroom Practice* are represented on a continuum and must be understood this way, not as a dichotomy. In focusing on improving learning opportunities for students and moving to the right on any of the *Shifts*, we recognize teaching as a complex and intellectually stimulating endeavor as we seek to figure out what specific actions might move us along the continuum. The Teaching Practices and *Shifts in Classroom Practice* can also be synthesized into look fors related to teacher actions. In the Appendix, you will find the Teaching Practices & *Shifts in Classroom Practice* Look Fors Bookmark, providing an at-a-glance resource for **professional learning**.

---

online resources ⌘ You can also download the bookmark at **resources.corwin.com/mathematicscoaching.**

Any one of these (interrelated) *Shifts*, or a combination of them, can be the focus of professional development, coaching conversations, and data gathering. The classroom practices described on the right side of the *Shifts*—whereby students are encouraged to understand, to reason, to hear, and to respond to ideas presented by their peers, and whereby the teacher challenges and supports their efforts to solve **high-level** mathematical **tasks**—are necessary in providing optimal learning opportunities for students to become mathematically proficient.

Connecting the *Shifts* to the Mathematical Practices

Imagine mapping the *Shifts* in Figure 1.4 to the Mathematical Practices in Figure 1.3 (or use the two Look Fors bookmarks). Identify one Mathematical Practice and see which *Shifts in Classroom Practice* you think would support student development of that Mathematical Practice. Did you identify one *Shift*? Three *Shifts*? Or did you see potential in each *Shift*? All of these responses could be considered correct. Work on any one *Shift* can affect student learning, as can work across the *Shifts*. Now imagine that you have done this same activity for all the Mathematical Practices. You will have a complex mapping of Mathematical Practices to *Shifts*. You can see that to develop mathematical proficiency for students, *all* the *Shifts in Classroom Practice* matter. And *any one* of the *Shifts in Classroom Practice* can contribute to student development of *any number* of the Mathematical Practices. In Chapter 12 of this book, we provide several professional development activities for engaging teachers in making these connections.

There are numerous ways to use the *Shifts* as tools to focus professional learning. You may decide to have your entire group of teachers from a particular setting (e.g., a course, grade, school, or entire district) select and focus on one *Shift* and have that be at the center of curriculum/lesson design, lesson study, or coaching cycles. Instead, or in addition, you may want to engage teachers in identifying their own *Shift* independently, selecting one that they feel will make the most difference in their own students' learning or one they feel is most needed to "move to the right." Instead of selecting a *Shift*, you may wish to have teachers take on all the *Shifts*, assessing where they are (see **Self-Assessment** Tool 2.1) and moving forward across the *Shifts*. A hybrid of these ideas is to select a subset of the *Shifts* that is most closely connected to a particular focus (e.g., **formative assessment**) and to work on that subset of *Shifts* (see a self-assessment tool as the first tool in each of the Focus Zone chapters).

Focus Zones

The *Shifts* themselves are multifaceted and complex. It may not be clear how these comprehensive teaching practices address needs for students or challenges in a particular setting. This raises the following question:

> *How do we (as coaches and teacher leaders) help teachers make the* Shifts in Classroom Practice *that lead to mathematical proficiency?*

When it comes to professional learning, it can be helpful to find a zone in which to work—one that addresses immediate needs of teachers and/or significant needs within a school or district setting. Generically, a *zone* is a separate area with a particular function, and that is exactly how we use the term here related to professional learning about mathematics teaching. A particular function might be to engage students or to analyze student work. Focusing on a zone provides an opportunity for pragmatic discussion, learning, and documenting of zone-specific strategies, ideas, and practices that can then be connected to the LMP Framework, with an eye constantly on developing mathematically proficient students.

There is certainly a myriad of possible **focus zones**. In this "everything-you-need book," we could not include every possibility! We selected eight Focus Zones based on these criteria:

- Is commonly encountered by mathematics coaches (based on surveys and input from you, the mathematics coaches)
- Has a research basis connecting the Focus Zone to student learning
- Has the greatest potential to shift classroom practices to the right

Our selected Focus Zones are listed in Figure 1.5. Each of these Focus Zones is a chapter in this book. And in that chapter is a Coach's Digest with resources for you and your teachers, as well as a Coach's Toolkit, which is a set of seven to ten tools to support professional learning and coaching cycles specific to that Focus Zone.

| Figure 1.5 | Focus Zones for Mathematics Professional Learning |

Chapter	Focus Zone
Chapter 3	Content Knowledge and Worthwhile Tasks
Chapter 4	Engaging Students
Chapter 5	Questioning and Discourse
Chapter 6	Formative Assessment
Chapter 7	Analyzing Student Work
Chapter 8	Differentiating Instruction for All Students
Chapter 9	Supporting Emergent Multilingual Students
Chapter 10	Supporting Students With Special Needs

Connecting Focus Zones to the LMP

Identification of a Focus Zone might occur in a variety of ways. As part of goal setting for the year, a group of teachers may select a focus (based on their self-assessment of the *Shifts in Classroom Practice*, for example). Or this could be part of a one-on-one coaching conversation—for example, with a beginning teacher. Together, you may agree that the focus of your work together will be on questioning and discourse. You both select this Zone because you see it as a way to work on *Shift 4* (from *show-and-tell* toward *share-and-compare*) or *Shift 6* (from *teaching so that students replicate procedures* toward *teaching so that students select efficient strategies*). In both cases, the *Shifts* were selected in order to develop one or more of the Mathematical Practices, such as Mathematical Practice 1 or both Mathematical Practices 1 and 3.

Enacting the Framework

We have described the LMP Framework in the way it can be used in designing professional learning experiences, beginning with the goal of student outcomes. Once the design is in place, it is essential to revisit the connections in the Framework (see the arrows on the right side in Figure 1.1). The selected Focus Zones support the *Shifts in Classroom Practice*.

Seeing how changes in a specific Zone, such as content and worthwhile tasks, are impacting several *Shifts* honors the efforts of teachers. Most importantly, it is crucial for teachers to see how efforts in a Focus Zone and/or on a *Shift in Classroom Practice* increase students' opportunities to learn (OTL). For example, as teachers use questioning and discourse (Focus Zone 5) to shift toward "teaching so that students select efficient strategies" (*Shift 6*), they need to see the different ways that students are developing flexibility and **efficiency** and the related Mathematical Practices that are therefore now evident in their students. It is very powerful to see how a particular instructional move or new routine can make significant changes in students' opportunities to learn important mathematics!

As you work on these three components of the LMP Framework, you may be leading professional learning and/or participating in coaching cycles. Professional learning ideas are provided in Chapters 2 through 10. Chapter 12 is focused specifically on presenting professional development, and Chapter 13 is focused on facilitating professional learning. The coaching cycle can play a critical role in professional learning, so it is important to consider some basic ideas to ensure coaching cycles are effective.

Coaching Cycle

The **coaching cycle** is commonly presented as a three-phase process: *pre-observation, observation,* and *post-observation*. This cycle and the titles of each component originated from clinical supervision models (Glickman, Gordon, & Ross-Gordon, 2001; Goldhammer, 1969). When this clinical supervision model is used in educational settings, it is typically incorporated as part of an evaluation system. **Coaching**, however, is not about evaluation; it is about learning. Therefore, we believe that mathematics coaches should not be involved in formal teacher evaluation. So we will adapt the language of the coaching cycle to focus on the collaborative activity in which the coach and teacher engage during each phase of the cycle and use the terms *plan, gather data,* and *reflect* (see Figure 1.6).

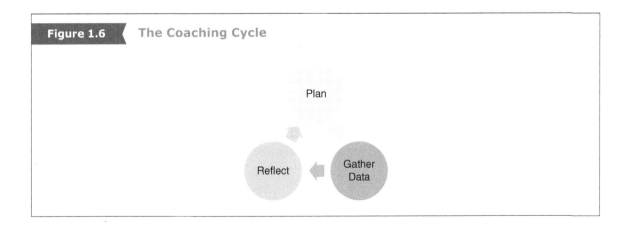

Figure 1.6 The Coaching Cycle

Plan

Reflect

Gather Data

The coaching cycle is dynamic and contextual. By *dynamic*, we mean that the cycle is flexible. You can begin at any phase of the coaching cycle. Of course, the ideal situation is that you have enough time in your day to conduct a complete coaching cycle with every teacher with whom you work! Obviously, that won't happen. But that doesn't mean that you then abandon the coaching cycle. Even if you do not have a planning conversation with a teacher or are not present during the lesson to collect any data, you can still have a reflecting conversation with the teacher that will support his or her growth. A teacher could ask you to come to her or his class and collect some specific data during a lesson. You could do this without previously having a planning conversation about the lesson. If you do not have the luxury of the time needed to complete the entire coaching cycle, engaging a teacher in any one part of the cycle is a valuable learning experience. By *contextual*, we mean that the implementation of the coaching cycle will be influenced by many factors, such as the teacher and coach's relationship, their beliefs about teaching and learning, and their educational experiences.

Phases of the Coaching Cycle

In this section, we briefly share the purpose of each phase, effective practice within each phase, and tips for using this book to support a coaching cycle.

Plan

Your goal during the planning phase of the coaching cycle is to support the teacher in effective lesson planning. This action may vary, depending on the teacher with whom you are working and the focus of the conversation. For example, you might be working with a beginning teacher who needs support in selecting goals and aligning lesson activities, or you might be working with a veteran teacher who is striving to engage all learners. Regardless of the specific situation, the coaching relationship should be collegial and reciprocal, one in which both the coach and the teacher are full participants, each learning from the discussion. Planning tools are in each of Chapters 2 through 10. Each tool provides "instructions to the coach" and general instructions as needed. The downloadable version of the tool does not include the "instructions to the coach," so it can be e-mailed to teachers or printed for them.

Gather Data

During this next phase of the coaching cycle, the coach is collecting data for the teacher. Thus, you and the teacher *together* decide what data will be gathered and what type of tool you will use for the data collection (see Coach's Toolkit in Chapters 2–10 for possibilities). The key to effective data collection is setting aside judgment and only collecting observable data (what you see and hear). For example, consider these two comments recorded on a data-gathering tool:

> *"Teacher asks **open-ended** question."*
> *"Teacher asks students to explain how they solved Problem 4."*

The first statement is a judgment about the type of question being asked. The latter statement is the actual data. Engaging the teacher in making judgments about the data collected can take place in the *Reflect* phase of the coaching cycle. You will notice that many of the data-gathering tools in the Coach's Toolkits provide significant space for recording data and then ways to later code that data *with* the teacher. The tools can be copied for teachers but are also are available for download so you can type data on the tool.

Reflect

Planning and data gathering are important, but their potential impact is realized through reflecting on the lesson and data. When we take the time to process our experiences, we gain insights that are essential to our professional growth. It is during the *Reflect* phase that any data gathered for the teacher is shared for discussion and analysis. It is also possible to have a reflecting conversation without any data to analyze. In either case, the questions you ask will support the teacher in reflecting on the lesson in critical ways. As with planning and data gathering, Chapters 2 through 10 provide options for reflecting.

Having tools to support each phase of the coaching cycle can help keep the focus on its intended goal. However, while we have placed tools under the headers of *Self-Assess, Plan, Gather Data,* and *Reflect,* many of the tools can be used in other phases. In the many stories we have heard from coaches, we often hear of only a single tool being used across the coaching cycle—for example, the coach uses the self-assessment tool but gathers evidence on sticky notes. These tools are meant to be a menu—select any that make sense for your setting/context.

Navigating the Coaching Cycle

Of course, simply understanding and having tools for the coaching cycle is not enough! In Chapter 11, we describe coaching skills that are critical for all coaching interactions, such as building trust and rapport, listening, paraphrasing, and posing questions. For each skill, we present a synopsis that includes a description of the skill, tips for effectively implementing the skill, insights from the field, and additional resources for further study. We recognize that these skills are complex, and becoming proficient in them is an ongoing process that develops over time. The briefs provided in Chapter 11 are intended to be a *Reader's Digest* version—at-a-glance support, reinforcement, and links to other resources that can support your work as a mathematics coach.

Getting Started

As you engage in coaching, we hope the LMP Framework provides a road map for you and your teachers. The Framework components place the purpose for change on goals for students (Mathematical Practices) via *Shifts in Classroom Practice* that can occur when professional development efforts focus on particular Focus Zones. As a coach, you can help teachers make strategic decisions on where to focus, as well as make explicit the connections between the components of the Framework (see Figure 1.2 and Tool 2.3). Your collaborative decision-making supports and is supported by the coaching cycle, as well as other professional learning opportunities.

Chapter 2 begins with an overview of the components of the LMP Framework already described, but it has been written for a teacher or administrator audience. This can be used to begin the conversation with teachers about setting goals related to mathematics learning and teaching. In your busy, multifaceted efforts to support teachers and their students, we truly hope that this book provides *almost* everything you need—at least as a starting place—because we recognize that your efforts as a mathematics coach are critical to improving mathematical learning opportunities for all students.

Chapter 2

Implementing Effective Teaching

Dear Coach,

Here is support for you as you work with teachers on implementing effective teaching.

In the **COACH'S DIGEST** ...

Overview: To review effective teaching and/or download and share with teachers.

Coaching Considerations for Professional Learning: Ideas for how to support a teacher or a group of teachers in learning about effective teaching, Mathematical Practices, and *Shifts in Classroom Practice.*

Coaching Lessons From the Field: Story or completed tool from mathematics coaches related to ideas from this chapter.

Coaching Questions for Discussion: Menu of prompts for professional learning or one-on-one coaching about the *Shifts* and Mathematical Practices.

Where to Learn More: Articles, books, and online resources for you and your teachers!

In the **COACH'S TOOLKIT** ...

Eleven tools focused on effective teaching, for professional learning or coaching cycles.

Coach's Digest

In the Coach's Digest, we begin with an overview of effective teaching, written to teachers (and to you, the coach). As you read the Overview, the following questions might help you reflect on this topic in terms of your role as a mathematics coach:

- Which of the Effective Mathematics Teaching Practices (and related *Shifts in Classroom Practice*) do you anticipate your teachers will want to prioritize?
- Which ones do you think need to be prioritized?

Overview of Implementing Effective Teaching

To download the Chapter 2 Overview, go to **resources.corwin.com/mathematicscoaching.**

Every student must have access to a high-quality mathematics learning experience. What does a high-quality learning experience look like? The answer is both simple and complex. Put simply, it is a daily experience in which a major focus is on mathematical proficiency, and therefore, the development of Mathematical Practices supersedes and interweaves with the content goals of a lesson. In other words, it matters *at least as much* that students can reason abstractly and quantitatively as that they can find the sum of two values. Mathematical proficiencies or processes are described effectively in the Mathematical Practices (NGA & CCSSO, 2010). Too often, the discussion of what students need to learn gets sidetracked with a dichotomy-type focus—for example, some argue that students need to understand what they are doing, while others argue they need to be efficient at using skills. The research is solid, however, in asserting that both strong conceptual understanding and procedural skills are absolutely essential in developing mathematical proficiency. The Mathematical Practices encompass this inclusive and comprehensive focus on mathematics; therefore, they must be a primary focus in any discussion about what students need to know and be able to do.

Developing mathematical proficiency (i.e., the Mathematical Practices) can be accomplished when teachers require that students engage in such practices as the way in which they learn about mathematics. There is significant evidence pointing toward Teaching Practices that support the development of mathematical proficiency, and these are comprehensively described in *Principles to Actions: Ensuring Mathematical Success for All* as Effective Mathematics Teaching Practices (see Figure 2.1).

Figure 2.1   Effective Mathematics Teaching Practices *From Principles to Actions: Ensuring Mathematical Success for All*

Effective Mathematics Teaching Practices	
	Effective teaching of mathematics . . .
Establish mathematics goals to focus learning.	. . . establishes clear goals for the mathematics that students are learning, situates goals within learning progressions, and uses the goals to guide instructional decisions.
Implement tasks that promote reasoning and problem-solving.	. . . engages students in solving and discussing tasks that promote mathematical reasoning and problem-solving and allows multiple entry points and varied solution strategies.
Use and connect mathematical representations.	. . . engages students in making connections among mathematical representations to deepen understanding of mathematics concepts and procedures and as tools for problem-solving.
Facilitate meaningful mathematical discourse.	. . . facilitates discourse among students to build shared understanding of mathematical ideas by analyzing and comparing student approaches and arguments.
Pose purposeful questions.	. . . uses purposeful questions to assess and advance students' reasoning and sense making about important mathematical ideas and relationships.
Build procedural fluency from conceptual understanding.	. . . builds fluency with procedures on a foundation of conceptual understanding so that students, over time, become skillful in using procedures flexibly as they solve contextual and mathematical problems.

Figure 2.1

	Effective teaching of mathematics ...
Support productive struggle in learning mathematics.	... consistently provides students, individually and collectively, with opportunities and supports to engage in productive struggle as they grapple with mathematical ideas and relationships.
Elicit and use evidence of student thinking.	... uses evidence of student thinking to assess progress toward mathematical understanding and to adjust instruction continually in ways that support and extend learning.

Source: National Council of Teachers of Mathematics (NCTM). (2014). Principles to Actions: Ensuring Mathematical Success for All. *Reston, VA: Author.*

Some of the Teaching Practices have clear and direct connections to Mathematical Practices. For example, supporting productive struggle (Teaching Practice) will support students' perseverance (Mathematical Practice). The better a teacher becomes at improving his or her ability to support productive struggle, the more a student has opportunities to make sense of mathematics and persevere. The effectiveness of a teacher in supporting productive struggle can be visualized as a continuum, as illustrated in Figure 2.2.

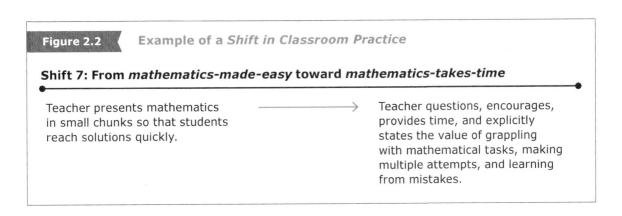

Figure 2.2 **Example of a *Shift in Classroom Practice***

Shift 7: From *mathematics-made-easy* toward *mathematics-takes-time*

Teacher presents mathematics in small chunks so that students reach solutions quickly. → Teacher questions, encourages, provides time, and explicitly states the value of grappling with mathematical tasks, making multiple attempts, and learning from mistakes.

Notice that the goal is to move Teaching Practices toward the right end of the continuum (which capture the essence of one of the NCTM Effective Mathematics Teaching Practices).

The connections between Mathematical Practices for students and Teaching Practices are actually not as straightforward or one-to-one as this example. Other *Shifts in Classroom Practice* can also impact a student's ability to make sense of and persevere in solving problems, for example. And teachers may not be focused specifically on a *Shift* or related teaching practice; instead, they may be zooming in on one topic—a Focus Zone such as formative assessment—and with this focus, their Teaching Practices are shifting to the right, thereby increasing students' opportunities to learn. This relationship is illustrated in the Leading for Mathematical Proficiency (LMP) Framework in Figure 2.3.

Professional learning can go in a myriad of directions. Attending a conference or reading a professional journal provides opportunities to explore many different zones. This can feel eclectic and unfocused, even overwhelming, as there are so many teaching ideas and **instructional strategies** that might support students. A way to bring cohesion to professional learning is to use this Framework. That means that whatever the professional learning focus might be, it is connected to the eight Effective Mathematics Teaching Practices, noticing professional growth along the *Shifts in Classroom Practice*, and ultimately, having data to support that this Practice or focus had an impact on students' emerging mathematical proficiency.

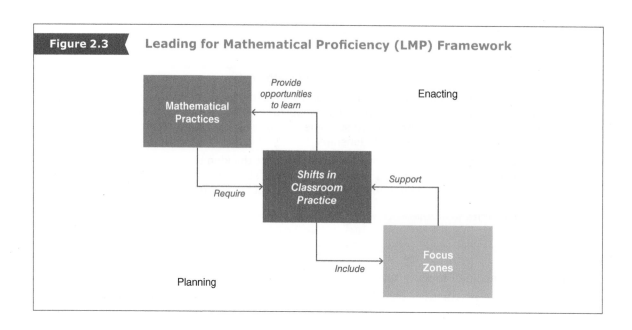

Figure 2.3 | **Leading for Mathematical Proficiency (LMP) Framework**

NOTES

_____
_____
_____
_____
_____
_____
_____
_____
_____
_____
_____
_____
_____
_____
_____
_____
_____
_____
_____
_____
_____
_____
_____
_____
_____
_____

Coaching Considerations for Professional Learning

One of the greatest challenges for a mathematics coach is to work with teachers to identify a focus for professional growth. This may be a short-term target for a coaching cycle or a long-term goal to be developed over the entire year. These goals will vary based on teachers' years of experience, as well as how receptive they are and how long you have been working with them. It is important to get a good feel for what strengths and needs a teacher has and to base the first experiences on this. Here are some ideas for how to engage in this work.

1. *Explore the Mathematical Practices.* There is a difference between having a general awareness of the standards versus having a deep understanding of what they mean. Time is well spent in providing opportunities for teachers to read and discuss aspects of the standards, considering what is new in terms of content and what this means for teaching. In particular, the Mathematical Practices require significant attention. A single Practice could be the focus of a workshop with teachers or a planning conversation. Teachers may identify those that they feel are (1) most difficult for them to get students to do and/or (2) most important for students at their grade or course level. This can be a good place to start. We provide a collection of professional development tools in Chapter 12 related to the Mathematical Practices. Additionally, several of the tools focus on any or all of the Mathematical Practices (see Tools 2.4, 2.6, and 2.9), and other tools in the Coach's Toolkit integrate the Mathematical Practices with effective planning and teaching (see Tools 2.3 and 2.5).

2. *Explore* **Shifts in Classroom Practice**. Like the Mathematical Practices, each of the *Shifts* is complex and must first be understood. Brainstorming teacher actions (things a teacher might *say* and things a teacher might *do*) at either end of the continuum, and in the middle, can help to solidify the focus of the *Shift* (see Tool 12.11). Teachers can self-assess where they feel they are with respect to each *Shift*, and this can begin a conversation about which *Shift* they feel will best support student engagement with the Mathematical Practices (see Tool 2.1). Additionally, several tools in the Coach's Toolkit focus on *Shifts in Classroom Practice* (see Tools 2.4, 2.8, and 2.10).

3. *Review video or print cases.* **Cases** provide an opportunity to look into a classroom of someone unknown, and this helps to develop an understanding of what it means to analyze, discuss, and adapt our teaching skills. It also provides concrete examples of the Mathematical Practices and the *Shifts*. Print cases provide the advantage of identifying specific teacher moves that result in particular student actions; video cases provide the opportunity to see a classroom in action. In the resource section that follows, we share some sites for finding cases, and new cases and resources continue to emerge.

Coaching Lessons From the Field

We have used the LMP Framework and the *Shifts in Classroom Practice* in some of our projects. The LMP Framework afforded a lens for leadership teams to examine what support teachers needed to be able to provide opportunities for students to interact with the Mathematical Practices.

The *Shifts* provided an opportunity for teachers, principals, or leadership teams to reflect on where they were as individuals or collectively as a school. This often set the stage for schools to consider their yearlong school goals and plans. The selected *Shift* would be written into the plan with details about how they could push that classroom practice further to the right on the continuum. Focusing on the *Shifts* with leadership teams helped to build continuity within the project,

(Continued)

(Continued)

allowing all stakeholders to focus on the same *Shift*.

We also used the *Shifts* to view video clips through the lens of a specific *Shift*. This often involved creating a giant-sized version of the specific *Shift* and posting it in the classroom. We typically selected two to three classroom video clips to view and recorded evidence of the identified *Shift* in action on sticky notes. Participants then placed the sticky note on the continuum in regard to where they felt the instruction would fall.

We used different-colored notes to distinguish each classroom video clip. Figure 2.4 shows a visual of this activity on a whiteboard. This activity promoted great discussion in regard to the focus *Shift* and led teachers to reflect on changes they needed to consider in their own practice in order to move to the right on the continuum.

—Denise Porter, Director
School Support Services
University of Chicago STEM Education

| Figure 2.4 | Teachers Record Evidence From Three Videos, Post Evidence Along Selected *Shift* Continuum |

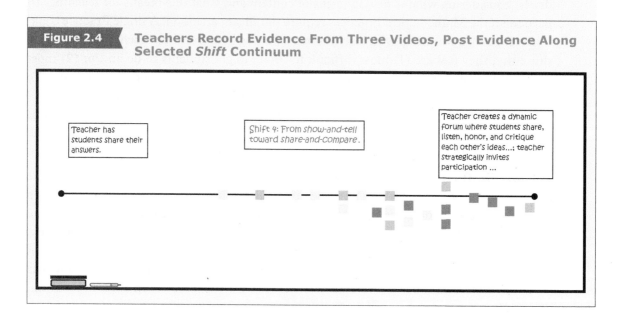

Teacher has students share their answers.

Shift 4: From *show-and-tell* toward *share-and-compare*.

Teacher creates a dynamic forum where students share, listen, honor, and critique each other's ideas...; teacher strategically invites participation ...

NOTES

_____

_____

_____

_____

_____

_____

_____

_____

_____

_____

_____

_____

_____

_____

_____

Questions Related to Implementing Effective Teaching

1. You may have noticed that the CCSS content standards are stated concisely and do not refer to other mathematics that is needed or precedes the stated standard. As you look at one standard, what do you do to determine how this individual standard connects to the other content the student has learned or will be learning?

2. Do you think that each of the Mathematical Practices is created equal? In other words, in your grade or course level which of these Mathematical Practices might be more applicable and/or require more explicit development on your part?

3. Are all the Teaching Practices and/or *Shifts in Classroom Practice* created equal? In other words, in your grade or course level which of the *Shifts* might be more applicable to your classroom?

Questions Related to the LMP Framework

At the heart of the Framework are the *Shifts in Classroom Practice*. Each is listed here with possible discussion questions.

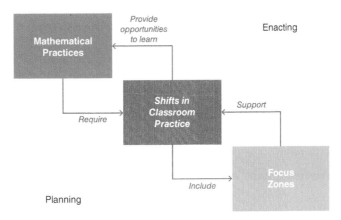

Shift 1: From *stating-a-standard* toward *communicating expectations for learning*

- What are some statements you might use to communicate expectations for learning?
- When and how might you communicate lesson goals in ways that make sense to students?

Shift 2: From *routine tasks* toward *reasoning tasks*

- What might be some ways to turn a routine task into a reasoning task?
- What might get in the way of students reasoning and thinking for themselves?
- How can you communicate to students that their reasoning is valued and important?

Shift 3: From *teaching about representations* toward *teaching through representations*

- What kind of support might students need in order to determine for themselves the correctness of their thinking?

- What are some ways that you make connections between mathematical content?
- How might you structure a lesson to use representations to strengthen mathematical understanding?

Shift 4: From *show-and-tell* toward *share-and-compare*

- What might be some benefits for students in making connections between mathematical ideas? For you, the teacher?
- What kind of supports might students need in order to learn how to explain their thinking to others?

Shift 5: From *questions that seek expected answers* toward *questions that illuminate and deepen student understanding*

- What kinds of instructional strategies might you use to include high-level thinking and complex tasks?
- In what ways might you ensure that all students have access and opportunity to engage in solving problems?
- What kind of support might students need in order to determine for themselves the correctness of their thinking?

Shift 6: From *teaching so that students replicate procedures* toward *teaching so that students select efficient strategies*

- How might you help students connect conceptual and procedural knowledge?
- How might you incorporate flexibility and strategy selection into lessons?
- How might you structure a lesson or task so that students have opportunities to contrast strategies?

Shift 7: From *mathematics-made-easy* toward *mathematics-takes-time*

- How might you provide structure or guidance for a task without giving away exactly how to do it?
- What might be some benefits of allowing students to engage in productive struggle?
- What are some ways that you challenge students to persevere on mathematical tasks?

Shift 8: From *looking at correct answers* toward *looking for students' thinking*

- How do you decide what representations, solution strategies, or concepts you will be looking for as you observe students working?
- What might be some ways you might gather data on student thinking during a lesson? As a written work sample?

Connecting *Shifts* to Mathematical Practices

- As you enact the ideas of Effective Mathematics Teaching, what do you see as changes in students' opportunities to learn?
- As you enact the ideas of Effective Mathematics Teaching, what do you see as ways students are developing proficiency with the Mathematical Practices?

Where to Learn More

Books

National Council of Teachers of Mathematics (NCTM). (2014). *Principles to Actions: Ensuring Mathematical Success for All.* Reston, VA: Author.

> *This most recent of NCTM's standards documents provides excellent, pragmatic, and research-based quick reads about each Teaching Practice, as well as Principles of Effective Teaching.*

National Council of Teachers of Mathematics (NCTM). (2017). *Taking Action: Implementing Effective Mathematics Teaching Practices.* Reston, VA: Author.

> *Like the previous book, these books (K–5, 6–8 and 9–12) are an elaboration of PtA. Each book focuses on the Teaching Practices, including* **vignettes***, tools for analyzing teaching and learning, and ideas for how to implement the ideas in the classroom. Good resources for PLCs, a book study, or other professional learning.*

Spangler, D. A., & Wanko, J. J. (Eds.). (2017). *Enhancing Classroom Practice With Research Behind Principles to Actions.* Reston, VA: NCTM.

> *This is one of several resources that followed the publication of PtA in order to provide more support to teachers and teacher leaders. This book has one chapter per Teaching Practice or Principle—each one provides a strong research base and pragmatic strategies for implementing that particular practice or principle. A good book for your own support as a coach or to study any one of these topics as a group.*

Van de Walle, J. A., Karp, K. S., & Bay-Williams, J. M. (2019). *Elementary and Middle School Mathematics: Teaching Developmentally* (10th ed.). Boston, MA: Pearson. [And the related *Teaching Student Centered Mathematics* three-book series, 2018.]

> *This comprehensive K–8 book is a go-to reference book. Part I of the book offers many good ideas and resources related to the standards and the Shifts. Part II addresses content, focusing on what is important and how to teach it. This book was also adapted into grade-band books (K–2, 3–5, 6–8)—for example, Teaching Student-Centered Mathematics: Grades 6–8 (Van de Walle, Lovin, Bay-Williams, & Karp, 2018). Each grade-band book provides additional activities and more elaboration on content for that grade band.*

Articles

Billings, E. M. H. (2017). "SMP That Help Foster Algebraic Thinking." *Teaching Children Mathematics, 23*(8), 476–483.

> *Algebraic thinking is something we can infuse into much of mathematics. This article provides instructional strategies and excellent question frames to help teachers infuse algebraic thinking, and therefore the Mathematical Practices. While this was written for K–5 teachers, it certainly applies across K–12.*

Bleiler, S. K., Baxter, W. A., Stephens, D. C., & Barlow, A. T. (2015). "Constructing Meaning: Standards for Mathematical Practice." *Teaching Children Mathematics, 21*(6), 337–344.

> *Wondering how to dig deeper into the meaning of the Mathematical Practices? This article describes tasks and activities teachers did in order to really make sense of the subtle nuances and complex ideas within a Mathematical Practice.*

Bostic, J. D., & Matney, G. T. (2014, Summer). "Role-Playing the Standards for Mathematical Practice: A Professional Development Tool." *NCSM Journal*, 3–10.

> *This article offers enough guidance that you can implement this professional learning with your teachers and provides great ideas for seeing what the Mathematical Practices look like in action.*

Mateas, V. (2016). "Debunking the Myths About the Standards for Mathematical Practice." *Mathematics Teaching in the Middle School*, 22(2), 92–99.

> *These myths are likely to resonate with you and to help you think about the Mathematical Practices, and this article can prompt rich discussion about how to effectively incorporate Mathematical Practices in the classroom. The examples are very helpful in illuminating key ideas about the Mathematical Practices. Access online at http://www.nctm.org/Publications/Mathematics-Teaching-in-Middle-School/2016/ Vol22/Issue2/Debunking-Myths-about-the-Standards-for-Mathematical-Practice.*

Online Resources

Illustrating the Standards for Mathematical Practice

http://www.mathedleadership.org/ccss/itp/index.html

> *Offering ready-to-use PowerPoints and activity pages focused on the Mathematical Practices, this site can be very useful for helping teachers see the Mathematical Practices as they connect to content at their grade level or for their course.*

Illustrative Mathematics

https://www.illustrativemathematics.org

> *Developed to provide support for teachers implementing CCSS, this site has a wealth of resources. You can select Mathematical Practices, identify the particular practice of interest, and find various vignettes and videos. They have also developed free curriculum that follows the progressions in the standards.*

Implementing the Mathematical Practice Standards (EDC)

http://mathpractices.edc.org

> *The Education Development Center (EDC) has a collection of resources for the mathematics coach for each of the Mathematical Practices. Examples include a mathematics task, connections to content standards, student dialogue for that task, teacher reflection questions, and more.*

Inside Mathematics

http://www.insidemathematics.org/common-core-resources/mathematical-practice-standards

> *Video excerpts of lessons involving teachers and students across elementary, middle, and high school grade levels as they are engaged in mathematics learning experiences help to illustrate the Mathematical Practices in action. Just as with the content standards, not every lesson reflects all elements of the individual Standards for Mathematical Practice. Videos highlight the many different ways teachers may promote student implementation of the Mathematical Practices in their classrooms with their learners.*

NCTM *Principles to Action* Professional Learning Toolkit

http://www.nctm.org/PtAToolkit

> *Professional learning modules have been created for each of the Teaching Practices that include mathematical tasks, narrative and video cases, student work samples, and vignettes. A wealth of information for a mathematics coach.*

Coach's Toolkit

These tools are a menu from which you can select any that make sense for your setting/context. They can be used independently or as part of a coaching cycle. You may start with the self-assessment, which can guide you in deciding which of the other tools may be most useful. If using these tools for a coaching cycle, mix and match as you like or use one of the combinations we suggest in the diagrams that follow. The tools in this chapter include instructions to the coach and the teacher. You can download copies of the tools that only have instructions for the teacher at **resources.corwin.com/mathematicscoaching**.

Self-Assess

2.1   *Shifts in Classroom Practice* Self-Assessment

Gather Data

2.6   Mathematical Practice Look Fors

2.7   *Shifts in Classroom Practice*

2.8   Effective Teaching Look Fors

Plan

2.2   Essential Planning Questions for Effective Teaching

2.3   Practices, *Shifts*, and Zones (Oh My)

2.4   Mathematical Practices by Design

2.5   Lesson Plan Template

Reflect

2.9   Noticing Mathematical Practices

2.10   Mapping Teaching Moves to *Shifts in Classroom Practice*

2.11   Effective Teaching of Mathematics

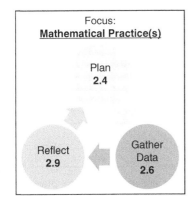

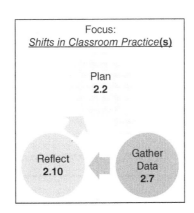

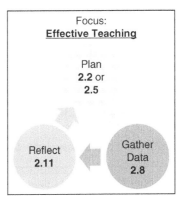

Additional Tools in Other Chapters

More focus on Mathematical Practices and *Shifts in Classroom Practice* can be found in all the Focus Zone chapters (Chapters 3–10).

online resources   To download the coaching tools for Chapter 2 that only have instructions for the teacher, go to **resources.corwin.com/mathematicscoaching**.

2.1 *Shifts in Classroom Practice* Self-Assessment

Instructions to Coach: Ask teachers (individually or as part of a PLC activity) to self-assess where they position themselves on each of these Shifts in Classroom Practice. *Use the results to focus on next steps. A one-page version of this tool without this note is available online for download.*

Instructions: Place an *X* along each continuum that best represents your classroom practice.

Shift 1: From *stating-a-standard* toward *communicating expectations for learning*

Teacher shares broad performance goals and/or those provided in standards or curriculum documents. → Teacher creates lesson-specific learning goals and communicates these goals at critical times within the lesson to ensure students understand the lesson's purpose and what is expected of them.

Shift 2: From *routine tasks* toward *reasoning tasks*

Teacher uses tasks involving recall of previously learned facts, rules, or definitions and provides students with specific strategies to follow. → Teacher uses tasks that lend themselves to multiple representations, strategies, or pathways encouraging student explanation (how) and justification (why/when) of solution strategies.

Shift 3: From *teaching about representations* toward *teaching through representations*

Teacher shows students how to create a representation (e.g., a graph or picture). → Teacher uses lesson goals to determine whether to highlight particular representations or to have students select a representation; in both cases, teacher provides opportunities for students to compare different representations and how they connect to key mathematical concepts.

Shift 4: From *show-and-tell* toward *share-and-compare*

Teacher has students share their answers. → Teacher creates a dynamic forum where students share, listen, honor, and critique each other's ideas to clarify and deepen mathematical understandings and language; teacher strategically invites participation in ways that facilitate mathematical connections.

Shift 5: From *questions that seek expected answers* toward *questions that illuminate and deepen student understanding*

Teacher poses closed and/or low-level questions, confirms correctness of responses, and provides little or no opportunity for students to explain their thinking. ⟶ Teacher poses questions that advance student thinking, deepen students' understanding, make the mathematics more visible, provide insights into student reasoning, and promote meaningful reflection.

Shift 6: From *teaching so that students replicate procedures* toward *teaching so that students select efficient strategies*

Teacher approaches facts and procedures with the goal of speed and accuracy. ⟶ Teacher provides time for students to engage with mathematical problems, developing flexibility by encouraging student selection and use of efficient strategies; teacher provides opportunities for students to evaluate when a strategy is best suited for the problem at hand.

Shift 7: From *mathematics-made-easy* toward *mathematics-takes-time*

Teacher presents mathematics in small chunks so that students reach solutions quickly. ⟶ Teacher questions, encourages, provides time, and explicitly states the value of grappling with mathematical tasks, making multiple attempts, and learning from mistakes.

Shift 8: From *looking at correct answers* toward *looking for students' thinking*

Teacher attends to whether an answer or procedure is (or is not) correct. ⟶ Teacher identifies specific strategies or representations that are important to notice; strategically uses observations, student responses to questions, and written work to determine what students understand; and uses these data to inform in-the-moment discourse and future lessons.

2.2 Essential Planning Questions for Effective Teaching

Instructions to Coach: This tool can be fully or partially filled out; a teacher can complete it prior to a planning conversation, or it can be the product of a planning conversation.

Instructions: The following questions can be part of a planning conversation or collaborative lesson design opportunity focused on implementing the Teaching Practices (TPs).

Teaching Practice Planning Questions	Lesson Planning Decisions
TP#1: What should students know (content) and be able to do (content and Mathematical Practices) in this lesson, and how will I make these expectations clear to students?	
TP#2: How will the task or lesson I selected include multiple pathways and elicit student reasoning?	
TP#3: What representations need to be connected to the content of this lesson, and does this mean I will select or that students will select the representations?	
TP#4: How and when will we discuss the important mathematical ideas of the lesson?	
TP#5: What questions will I ask to help students make connections among solution strategies and among mathematical ideas?	
TP#6: How will I ensure students are understanding strategies/algorithms and learning to make good choices about when to use a particular strategy/algorithm?	
TP#7: What structures, feedback, and instructional moves can I use to ensure that students grapple with the tasks in the lesson and strive to make mathematical connections?	
TP#8: What will I have students do at the very end of the lesson that summarizes the big ideas of the lesson?	

2.3 Practices, *Shifts*, and Zones (Oh My)

Instructions to Coach: You can identify a starting place or ask teacher(s) to select one. As they work, focus on evidence of what it looks like in practice.

Instructions: Identify the Mathematical Practices, *Shifts*, and Focus Zones that will be the target of your lesson/unit. You may start in any of the boxes, discussing how selections in one area connect to selections in the other areas.

	Selection and what it looks like
Mathematical Practices 1. Make sense of problems and persevere in solving them. 2. Reason abstractly and quantitatively. 3. Construct viable arguments and critique the reasoning of others. 4. Model with mathematics. 5. Use appropriate tools strategically. 6. Attend to precision. 7. Look for and make use of structure. 8. Look for and express regularity in repeated reasoning.	
Shifts in Classroom Practice **...Toward...** 1. Communicates learning expectations. 2. Reasoning tasks. 3. Teaching through representations. 4. Share-and-compare. 5. Questions illuminate and deepen student understanding. 6. Selecting efficient strategies. 7. Mathematics-takes-time. 8. Valuing students' thinking.	
Focus Zones 1. Content Knowledge and Worthwhile Tasks 2. Engaging Students 3. Questioning and Discourse 4. Formative Assessment 5. Analyzing Student Work 6. Differentiating Instruction for All Learners 7. Supporting Emerging Multilingual Students 8. Supporting Students with Special Needs	

2.4 Mathematical Practices by Design

Instructions to Coach: Not all Mathematical Practices fit all lessons. And even if many are likely to be evident, it is best to focus on fewer (1–3), and these should be based on "best fit" for the learning outcome and the task selected. Discuss which MPs make sense, and then discuss design plans.

Instructions: Highlight the Mathematical Practices and/or Look Fors that will be prominent in the lesson. Write design plans for how the selected Mathematical Practice will be developed.

Topic/Goal of Lesson: _____

Mathematical Practices and Student Look Fors	Design Plans
1. Make sense of problems and persevere in solving them. • Analyze information (givens, constraints, relationships, goals). • Make conjectures and plan a solution pathway. • Use objects, drawings, and diagrams to solve problems. • Monitor progress and change course as necessary. • Check answers to problems and ask, "Does this make sense?"	
2. Reason abstractly and quantitatively. • Make sense of quantities and relationships in problem situations. • Create a coherent representation of a problem. • Translate from contextualized to generalized or vice versa. • Flexibly use properties of operations.	
3. Construct viable arguments and critique the reasoning of others. • Make conjectures and use counterexamples to build a logical progression of statements to support ideas. • Use definitions and previously established results. • Listen to or read the arguments of others. • Ask probing questions to other students.	
4. Model with mathematics. • Determine equation that represents a situation. • Illustrate mathematical relationships using diagrams, two-way tables, graphs, flowcharts, and formulas. • Check to see whether an answer makes sense within the context of a situation and change a model when necessary.	

Mathematical Practices and Student Look Fors	Design Plans
5. Use appropriate tools strategically. • Choose tools that are appropriate for the task (e.g., manipulative, calculator, digital technology, ruler). • Use technological tools to visualize the results of assumptions, explore consequences, and compare predictions with data. • Identify relevant external math resources (digital content on a website) and use them to pose or solve problems.	
6. Attend to precision. • Communicate precisely, using appropriate terminology. • Specify units of measure and provide accurate labels on graphs. • Express numerical answers with appropriate degree of precision. • Provide carefully formulated explanations.	
7. Look for and make use of structure. • Notice patterns or structure, recognizing that quantities can be represented in different ways. • Use knowledge of properties to efficiently solve problems. • View complicated quantities both as single objects and as compositions of several objects.	
8. Look for and express regularity in repeated reasoning. • Notice repeated calculations and look for general methods and shortcuts. • Maintain oversight of the process while attending to the details. • Evaluate reasonableness of intermediate and final results.	

Source: Adapted from Elementary Mathematics Specialists & Teacher Leaders Project. (n.d.). Common Core Look-Fors. Unpublished document. Used with permission. Previously published by Bay-Williams, J., McGatha, M., Kobett, B., and Wray, J. (2014). *Mathematics Coaching: Resources and Tools for Coaches and Leaders, K–12.* New York, NY: Pearson Education, Inc.

2.5 Lesson Plan Template

Instructions to the Coach: This tool can be prepared by a teacher in advance or used in collaborative planning. Planning questions are provided within each cell. Use the "Lesson Reflections" cell to determine what data-gathering tool to use.

Instructions: Complete this tool in connection with a specific lesson and selected task.

Content standard(s): Objectives:	Mathematical Practices ☐ 1. Make sense of problems and persevere in solving them. ☐ 2. Reason abstractly and quantitatively. ☐ 3. Construct viable arguments and critique the reasoning of others. ☐ 4. Model with mathematics. ☐ 5. Use appropriate tools strategically. ☐ 6. Attend to precision. ☐ 7. Look for and make use of structure. ☐ 8. Look for and express regularity in repeated reasoning.
Essential questions: *What questions will promote inquiry, understanding, and transfer of learning?*	Assessment evidence: *By what criteria will "performance of understanding" be judged?*
Focus task: *What specific mathematical activities, investigations, texts, problems, or tasks will students do in order to learn the content?*	Anticipated student responses: *What prior knowledge or limited conceptions might students have? How might students solve the problem?*
Resources: *What materials or resources are essential for students to successfully complete the lesson tasks or activities?*	Anticipated language needs: *What words, phrases, or symbols may need to be explicitly discussed within the lesson?*

Engage (set up the task): *Exactly how will I elicit prior content knowledge, connect to students' experiences, and set up the task (to ensure students understand the task without overscaffolding or funneling)?*

Explore (solve the task): *What questions might I ask individuals or small groups of students that focus on the content and Mathematical Practices?*

Connect (discuss task and related mathematical concepts): *What questions and/or activity will engage students in explaining and/or illustrating the concepts of the lesson, as well as provide formative assessment as to who learned what?*

Lesson reflections: *What questions connected to the standards and assessment evidence will I use to reflect on the effectiveness of this lesson?*

2.6 Mathematical Practice Look Fors

Instructions to the Coach: You can fill out this tool during an observation and/or have the teacher complete it with what they are seeing and hearing. Or you can video or write down as much student talk as possible and map that data to this page in a reflecting conversation.

Instructions: During a lesson, listen for student actions related to any or all of these Mathematical Practices. Note what they said or did in the examples column.

Mathematical Practice	Student Look Fors	Examples
1. Make sense of problems and persevere in solving them.	☐ Analyze information (givens, constraints, relationships, goals). ☐ Make conjectures and plan a solution pathway. ☐ Use objects, drawings, and diagrams to solve problems. ☐ Monitor progress and change course as necessary. ☐ Check answers to problems and ask, "Does this make sense?"	
2. Reason abstractly and quantitatively.	☐ Make sense of quantities and relationships in problem situations. ☐ Create a coherent representation of a problem. ☐ Translate from contextualized to generalized or vice versa. ☐ Flexibly use properties of operations.	
3. Construct viable arguments and critique the reasoning of others.	☐ Make conjectures and use counterexamples to build a logical progression of statements to support ideas. ☐ Use definitions and previously established results. ☐ Listen to or read the arguments of others. ☐ Ask probing questions to other students.	
4. Model with mathematics.	☐ Determine equation that represents a situation. ☐ Illustrate mathematical relationships using diagrams, two-way tables, graphs, flowcharts, and formulas. ☐ Check to see whether an answer makes sense within the context of a situation and change a model when necessary.	

Mathematical Practice	Student Look Fors	Examples
5. Use appropriate tools strategically.	☐ Choose tools that are appropriate for the task (e.g., manipulative, calculator, digital technology, ruler). ☐ Use technological tools to visualize the results of assumptions, explore consequences, and compare predictions with data. ☐ Identify relevant external math resources (digital content on a website) and use them to pose or solve problems.	
6. Attend to precision.	☐ Communicate precisely using appropriate terminology. ☐ Specify units of measure and provide accurate labels on graphs. ☐ Express numerical answers with appropriate degree of precision. ☐ Provide carefully formulated explanations.	
7. Look for and make use of structure.	☐ Notice patterns or structure, recognizing that quantities can be represented in different ways. ☐ Use knowledge of properties to efficiently solve problems. ☐ View complicated quantities both as single objects and as compositions of several objects.	
8. Look for and express regularity in repeated reasoning.	☐ Notice repeated calculations and look for general methods and shortcuts. ☐ Maintain oversight of the process while attending to the details. ☐ Evaluate reasonableness of intermediate and final results.	

Source: Adapted from Elementary Mathematics Specialists & Teacher Leaders Project. (n.d.). Common Core Look-Fors. Unpublished document. Used with permission. Previously published by Bay-Williams, J., McGatha, M., Kobett, B., and Wray, J. (2014). *Mathematics Coaching: Resources and Tools for Coaches and Leaders, K–12.* New York, NY: Pearson Education, Inc.

2.7 Shifts in Classroom Practice

Instructions to the Coach: Record each teacher action and interaction in a separate box (copy as needed). At the end of the lesson, cut out each piece of evidence and use Tool 2.10 to determine where each piece belongs on the Shift's continuum. Alternatively, you might record each event on a sticky note.

2.8 Effective Teaching Look Fors

Instructions to the Coach: Preselect practices with a teacher and fill out this tool during an observation. Or video or write down as much teacher talk/actions as possible and map that data to this page in a reflecting conversation.

Instructions: Select Teaching Practice(s) and record specific teacher moves or actions that demonstrate that Practice.

Teaching Practice (NCTM, 2014) Look Fors	Evidence
Establish mathematics goals to focus learning. ☐ Goals are appropriate, challenging, and attainable. ☐ Goals are specific to the lesson and clear to students. ☐ Goals are connected to other mathematics. ☐ Goals are revisited throughout the lesson.	
Implement tasks that promote reasoning and problem-solving. ☐ Chooses engaging, high-cognitive-demand tasks with multiple solution pathways. ☐ Chooses tasks that arise from home, community, and society. ☐ Uses how, why, and when questions to prompt students to reflect on their reasoning.	
Use and connect mathematical representations. ☐ Uses tasks that lend themselves to multiple representations. ☐ Selects representations that bring new mathematical insights. ☐ Gives students time to select, use, and compare representations. ☐ Connects representations to mathematics concepts.	
Facilitate meaningful mathematical discourse. ☐ Helps students share, listen, honor, and critique each other's ideas. ☐ Helps students consider and discuss each other's thinking. ☐ Strategically sequences and uses student responses to highlight mathematical ideas and language.	

Teaching Practice (NCTM, 2014) Look Fors	Evidence
Pose purposeful questions. ☐ Questions make the mathematics visible. ☐ Questions solidify and extend student thinking. ☐ Questions elicit student comparison of ideas and strategies. ☐ Strategies are used to ensure every child is thinking of answers.	
Build procedural fluency from conceptual understanding. ☐ Gives students time to think about different ways to approach a problem. ☐ Encourages students to use their own strategies and methods. ☐ Asks students to compare different methods. ☐ Asks why a strategy is a good choice.	
Support productive struggle in learning mathematics. ☐ Provides ample wait time. ☐ Talks about the value of making multiple attempts and persistence. ☐ Facilitates discussion on mathematical error(s), misconception(s), or struggle(s) and how to overcome them.	
Elicit and use evidence of student thinking. ☐ Identifies strategies or representations that are important to look for as evidence of student understanding. ☐ Makes just-in-time decisions based on observations, student responses to questions, and written work. ☐ Uses questions or prompts that probe, scaffold, or extend students' understanding.	

Source: Previously published by Bay-Williams, J., McGatha, M., Kobett, B., and Wray, J. (2014). Mathematics Coaching: Resources and Tools for Coaches and Leaders, K–12. New York, NY: Pearson Education, Inc.

2.9 Noticing Mathematical Practices

Instructions to Coach: This tool can be used as a follow-up to Tools 2.3 Practices, *Shifts,* and Zones (Oh My) *and 2.6 Mathematical Practice Look Fors. Teachers can have it already filled out for a reflecting conversation, or you can use these as discussion prompts.*

Instructions: In reflecting on the lesson related to evidence of students engaged in using the Mathematical Practices, respond to the following questions.

1. What evidence do you see of one instructional move or *Shift* that supported the students engaging in the Mathematical Practices?

2. What aspects of the lesson (design, questions, grouping, task) do you think were particularly effective at eliciting these Mathematical Practices or Look Fors?

3. What aspect(s) of the Look Fors were not as evident in the lesson? What might be adapted in the lesson to better support the development of those Look Fors?

4. What might be one *Shift in Classroom Practice* that could be a target for future professional development that is needed in order to better engage in the identified Look Fors?

2.10 Mapping Teaching Moves to *Shifts in Classroom Practice*

Instructions to Coach: For professional learning, assign different groups to different Shifts and sort evidence from a classroom teaching video (see Tool 2.7 for recording evidence). For lesson cycle, use Tool 2.7 to gather data and then ask the teacher to place each teacher move along the continuum. Or you could each identify a location and then discuss any differences.

Selected *Shift*: _____

Cut along the dashed lines. Tape together to make a long continuum (e.g., on flip chart paper). Glue, tape, or write your selected *Shift* descriptors in the empty boxes. Decide where on the continuum the evidence might fit (and why).

2.11 Effective Teaching of Mathematics

Instructions to Coach: Select any of these prompts or use them all—the first six address the Teaching Practices, and the final two are summarizing prompts.

Instructions: Using data from a lesson, reflect on the questions below.

1. To what extent did the lesson tasks, activities, and/or discussion support the lesson objectives?

2. What representations/strategies/approaches did students use to solve problems and demonstrate their understanding? Which ones may need more attention going forward?

3. What questioning and discourse went well, and what was challenging?

4. In what ways did you see students making connections (between concepts and procedures; between representations; to previously learned concepts; to the real world)?

5. In what ways were students asked to grapple with mathematics? How effective were your strategies to support their struggle without taking away students' thinking?

6. To what extent were you able to determine whether each student learned the objectives?

7. What do you feel were the most successful aspects of this lesson?

8. What are you learning that you would like to remember when you teach this lesson again?

Part II

Exploring Zones on the Journey
Professional Learning Focus Areas

Now that you have had time to explore the Leading for Mathematical Proficiency (LMP) Framework, your journey is going to include attending to specific topics that support mathematics teaching! We call these identified focus areas *Focus Zones*. Coaching is most effective when we focus on particular kinds of professional learning with teachers because it helps teachers target specific teaching practices. In this part of the book, you will explore content and tools associated with (a) **content knowledge** and **worthwhile tasks**, (b) engaging students, (c) questioning and discourse, (d) formative assessment, (e) analyzing student work, (f) differentiating instruction for all students, (g) supporting emergent multilingual students, and (h) supporting students with special needs. With your teachers, select the chapter(s) in this section that are most relevant to your needs or that emerged as most important through self-assessment activities from Chapter 2.

Each chapter begins with a Coach's Digest in which we have tried to provide you *everything you need* to support teachers related to this zone—in 10 pages or less! This includes an overview that teachers might read, connections to *Shifts* and Mathematical Practices, discussion questions you could pose to teachers, professional development activities, and where to go to study the topic in more depth (books, articles, and online resources). *Everything you need* must include tools for coaching, so following the Coach's Digest are several tools to self-assess, plan, gather data, and reflect with teachers as part of professional development or coaching cycles.

Chapter 3

Content Knowledge and Worthwhile Tasks

Dear Coach:

Here is support for you as you work with teachers on content knowledge and worthwhile tasks.

In the **COACH'S DIGEST**...

Overview: Highlights about content knowledge and tasks for you to download and share with teachers.

Coaching Considerations for Professional Learning: Ideas for how to support a teacher or a group of teachers in deepening their content, as well as exploring content and worthwhile tasks for their students.

Coaching Lessons from the Field: Story from a mathematics leader sharing how she used tools from this chapter.

Connecting to the Framework: Specific ways to connect selected *Shifts* and Mathematical Practices to content and tasks.

Coaching Questions for Discussion: Menu of prompts for professional learning or one-on-one coaching about content knowledge and using worthwhile tasks.

Where to Learn More: Articles, books, and online resources for you and your teachers with more examples of content knowledge and worthwhile tasks!

In the **COACH'S TOOLKIT**...

Ten tools focused on content knowledge and worthwhile tasks, for professional learning or coaching cycles.

Coach's Digest

In the Coach's Digest, we begin with an overview of content knowledge and worthwhile tasks, written to teachers (and to you, the coach). As you read the Overview, the following questions might help you reflect on this topic in terms of your role as a mathematics coach:

- How might I use the Levels of Cognition to help teachers analyze tasks and instruction?
- How might I provide experiences for teachers in opening up and increasing the cognitive demand of tasks?

online resources 📥 To download the Chapter 3 Overview, go to **resources.corwin.com/mathematicscoaching.**

The mathematical knowledge of teachers impacts their students' achievement, in particular the **mathematical knowledge for teaching (MKT**; Hill, Rowan, & Ball, 2005). And this specialized knowledge develops over time as teachers analyze curriculum and possible learning trajectories for students, consider common errors or misconceptions, and implement worthwhile tasks. Therefore, time spent focused on developing mathematical knowledge connected to what a teacher is teaching is time well spent! While there are many ways to "get at" MKT, here we zoom in on two important ideas: (1) understanding the relationship between and the importance of conceptual and procedural knowledge and (2) considering what makes a task worthwhile and how to adapt a task to make it worthwhile.

Developing Fluency

The fact that "Build procedural fluency from conceptual understanding" is one of eight Teaching Practices is an indication of the critical importance of these two knowledge domains. Rather than pit procedural knowledge against conceptual knowledge, both can be thought of as on a continuum from weak to strong (Star, 2005). This is a much more useful way to present conceptual and procedural knowledge because, in fact, procedural fluency requires depth of knowledge in *both*. Too often, **fluency** instruction is limited to developing facility with a single procedure for a particular topic. But **procedural fluency** includes four components: accuracy, efficiency, appropriate **strategy** selection, and flexibility (NGA & CCSSO, 2010; NRC, 2001). These final two components are really key to actual fluency and must be included in fluency instruction. As illustrated in Figure 3.1, efficiency is not only about speed but also about strategy selection. For example, how might you efficiently solve the problem 3,005–1,998? A mental counting-up (or back) strategy is more efficient than applying the standard **algorithm** (that requires regrouping over zeros).

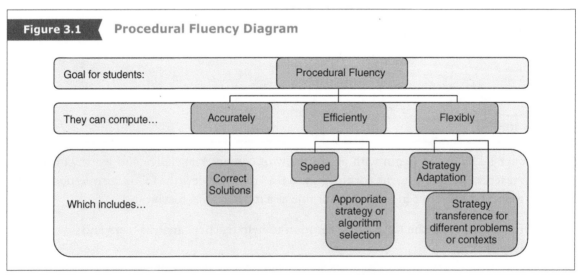

Figure 3.1 **Procedural Fluency Diagram**

Goal for students: → Procedural Fluency

They can compute… → Accurately | Efficiently | Flexibly

Which includes… → Correct Solutions | Speed / Appropriate strategy or algorithm selection | Strategy Adaptation / Strategy transference for different problems or contexts

Source: *Bay-Williams, J. M., and Stokes Levine, A. (2017). "The Role of Concepts and Procedures in Developing Fluency." In D. Spangler & J. Wanko (Eds.),* Enhancing Professional Practice With Research Behind Principles to Actions. *Reston, VA: NCTM.*

Strategy selection and flexibility require high-level thinking. Bloom's **Taxonomy** provides a useful framework to consider low-level to high-level thinking (see Chapter 5). Fan and Bokhove (2014) organized Bloom's Taxonomy and computational actions into three levels of cognition (see Figure 3.2). Level 1 computational actions are low level on Bloom's Taxonomy (*Remember*). Notice that these actions, however, have dominated mathematics teaching and learning, sometimes being the sole focus of worksheets, textbook lessons, and classroom discussions. Levels 2 and 3 are high-level computational actions. A key distinction between Levels 2 and 3 is that Level 2 focuses on understanding one particular procedure, and at Level 3, the focus is on contrasting and comparing several procedures (Fan & Bokhove, 2014).

Figure 3.2	Mapping Fluency Thinking to Bloom's (Revised) Taxonomy		
A lower level supports a higher level and vice versa.	**Cognitive Levels (Fan & Bokhove, 2014)**	**Bloom's (Revised) Elements**	**Related Actions With Procedures**
	Level 1. **Knowledge and Skills**	1. Remember	Tell the steps of a procedure. Carry out steps in a straightforward situation.
	Level 2. **Understanding and Comprehension**	2. Understand	Describe why a procedure works. Apply procedure to complex problems.
		3. Apply	
		4. Analyze	
	Level 3. **Evaluation and Construction**	5. Evaluate	Compare different algorithms. Judge efficiency of an algorithm. Construct new algorithms (strategies). Generalize when a procedure works.
		6. Create	

Source: *Based on Fan, L., and Bokhove, C. (2014). "Rethinking the Role of Algorithms in School Mathematics: A Conceptual Model With Focus on Cognitive Development." ZDM International Journal on Mathematics Education, 46(3), 481–492.*

Historically, school mathematics learning has had an overemphasis on demonstrating one procedure or algorithm for a particular problem type and having students practice it (Level 1). This results in weak procedural knowledge and weak conceptual knowledge. Mathematics tasks and classroom instruction must spend significant time on Levels 2 and 3 for students to develop procedural fluency. Consider how a topic such as division of fractions can look across these levels, using the example $4\frac{1}{2} \div \frac{3}{4}$. At Level 1, a student might be shown and asked to remember the "invert and multiply" algorithm. At Level 2, students are able to describe the meaning of the operation. A student might say, "I am trying to find how many $\frac{3}{4}$ are in $4\frac{1}{2}$," or might describe a situation, "This is like needing $\frac{3}{4}$ of a yard of fabric to make each apron, and finding how many aprons can be made from $4\frac{1}{2}$ yards of fabric." And they can describe why an algorithm works. For example, noting that the reason you multiply by 4 is that you can first count how many fourths in $4\frac{1}{2}$. Since each whole has 4 fourths, you multiply. And since you are finding groups of three (for $\frac{3}{4}$), you divide by 3. At Level 3, students are asked to consider various procedures and expected to flexibly select one that fits the numbers. In this case, they may invert and multiply, or they may use a counting-up strategy: Two aprons would take $1\frac{1}{2}$ yards, so four aprons would take 3 yards, so six aprons would take $4\frac{1}{2}$ yards. Or they may mentally think of $4\frac{1}{2}$ as 18 fourths and then divide 18 by 3 to equal 6. One way to focus on Level 3 is to analyze a worked example or compare two student-worked examples, a practice that can improve student achievement (Renkl, 2014; Star & Verschaffel, 2017). Worked examples can be correct, incorrect, or incomplete (i.e., a student gets "stuck"), and students are asked to analyze the strategy, find the error, or help complete the task. Notice how conceptual knowledge supports and connects to procedural knowledge when working at Levels 2 and 3 and that the result is a deeper understanding of the topic.

Worthwhile Tasks

The development of deep-content knowledge occurs when students have the opportunity to engage in worthwhile mathematics tasks—worthwhile because they provide students an opportunity to apply important mathematical properties, make connections, and think at a high level. To provide these opportunities to learn (OTL) requires finding such a task, then maintaining the rigor or high-level thinking of the task in planning and in teaching. This is effectively described in the Mathematical Tasks Framework (Smith & Stein, 1998; see Figure 3.3).

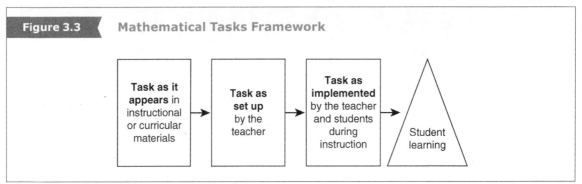

Figure 3.3   **Mathematical Tasks Framework**

The goal for mathematics teachers is to maintain the level of cognitive demand in a task rather than "help" students in ways that lower the level of demand. Numerous studies have found that increased exposure to cognitively challenging tasks and extended engagement with high-level cognitive demands increases students' learning of mathematics (e.g., Hiebert & Wearne, 1993; NCES, 2003; Stein & Lane, 1996).

To start, worthwhile tasks should feel worthwhile to a student. That means the context and/or the mathematical questions posed must be engaging. Relevant contexts serve the purpose of providing a concrete link to the abstract mathematical ideas, as well as seeing how mathematics can be used to learn about and to serve the school and/or community. Using everyday situations can increase student participation, increase student use of different problem strategies, and help students develop a productive disposition (Tomaz & David, 2015).

Second, worthwhile tasks also have **multiple entry points**, meaning that the task can be approached in a variety of ways and has varying degrees of challenge within it. Having multiple entry points serves various purposes. First and foremost, such tasks better meet the needs of diverse learners because students can select an approach based on their prior experiences and knowledge. Second, having multiple entry points opens up the opportunity to compare and evaluate strategies for solving the problem. This provides an opportunity to see relationships among representations, as well as see different ways to symbolically solve a problem and discuss efficient approaches. Third, having multiple entry points provides more insights into student thinking and understanding, providing important and useful formative assessment data.

Worksheets and textbook problems can often be closed, low level, and uninteresting, but there are often "tweaks" that can be made to the instructions or to a task that can change it into a much more interesting and higher-level thinking activity. Boaler (2016) provides suggestions on adapting procedural tasks with such a goal in mind. In addition to multiple entry points, these ideas include the following:

- *Grow the task:* Change a task from a single computation to finding possibilities. For example, rather than add 23 + 15, find numbers that result in 50.
- *Invite multiple ways:* Ask students to use multiple strategies and representations.
- *Add a visual requirement:* Ask students to show two different visuals or connect between representations (e.g., story and equation).
- *Reason and convince:* Ask students to create convincing arguments and to expect the same from their peers.

NOTES

_____

_____

_____

_____

_____

_____

_____

_____

_____

_____

_____

_____

_____

_____

_____

_____

_____

_____

_____

_____

_____

_____

_____

_____

_____

_____

_____

_____

Coaching Considerations for Professional Learning

Hopefully, the discussion to this point has communicated the critical importance of teachers thinking deeply about the content they are teaching. Here are some ways to do this related to content knowledge.

1. ***Address myths, misinterpretations, and overgeneralizations***. Developing a shared understanding of what the goals are in developing content is critical—yet there are many sound bites that can interfere with our shared understanding of developing content. Here are a few:

   a. *Conceptual knowledge is good, and procedural knowledge is bad (drill and kill, etc.)*. In fact, both are needed to ensure mathematical proficiency (NRC, 2001; Star, 2015). *Appropriate* balance and connections between them is critical. This must be our message if we are going to help teachers develop mathematically proficient students (see Tool 3.7, which can be used with a video or actual observation).

   b. *Procedural knowledge is equivalent to drill and practice.* Distinguishing between procedural knowledge and drill is important. Drill might lead to weak procedural knowledge, while experiences selecting and using strategies and learning why algorithms work can develop strong procedural knowledge. Drill can provide opportunities for students to practice strategy selection and flexibility, as can discussions of the problems. The Overview addresses this misconception (see also Tool 3.3).

   c. *Students naturally see the connection between concepts and procedures.* Too often, a topic is introduced with a tool or representation (e.g., number lines, base 10 blocks, fraction bars), and the next day the discussion moves to doing a related procedure, assuming the visuals provided previously were all that were needed. Students need regular opportunities to describe and illustrate the connections between visual and symbolic representations. Teachers can informally ask students as they are working to show how a procedure is connected to a tool or representation (see Tools 3.4 and 3.9, which focus on making connections between concepts and procedures and among representations).

   d. *Open-ended prompts are higher-level.* Consider the teacher question "How did you solve this problem?" If the teacher's intent is to hear if the student followed an algorithm correctly, this is a knowledge question (**low-level**). Helping teachers see what type of questions are higher-level can help them to increase the level of reasoning and sense making in their classrooms. Because higher-level questions focus on why and when a particular strategy, representation, or pathway works, shifting toward high-level questions moves the focus to procedural fluency and mathematical proficiency (see also Chapter 4).

   e. Hard *means cognitively demanding.* There is a difference between a problem that is very messy and a problem that is **cognitively demanding**. For example, consider the expression $2(3 + x) - x(1 - 3x) - 8$. It may be considered hard for various reasons (e.g., subtraction outside of the parentheses, negative x times a negative $3x$). But is it cognitively demanding? The answer lies in the extent to which students will be asked to make connections (in this case, to properties, to other problems like this one, to equivalence, to the meaning of the variable or operations, etc.; see Tools 3.2, 3.5, 3.8, and 3.10).

One idea is to jigsaw these statements (a–e) and ask groups of teachers to provide illustrations or non-examples as to why they are not accurate. This could also be paired with reading the Overview or a related article.

2. ***Support the creation of concept maps.*** In light of the emphasis on learning trajectories and progressions that was prompted by the CCSS-M, it is helpful to consider the content that gets learned before and after a particular idea. It is also important to see mathematics as an integrated body of knowledge rather than a set of topics, and the concept map accomplishes this as well. In fact, the great thing about engaging teachers in creating concept maps is that it helps them better understand the way curriculum is articulated and strengthens their content knowledge. Finally, and importantly, concept maps are excellent tools for students—and teachers will be more likely to use them with students if they have engaged in doing some themselves.

3. ***Analyze a task.*** Have teachers solve a task, anticipate student responses to that task, and then consider how they would set that task up to teach it in their classrooms. Sometimes, teachers suggest that much more support/input is needed in the setting up of the task, and focusing on whether this support maintains the level of cognitive demand can be a very valuable discussion (see Tool 3.5 or 3.6).

4. ***Evaluate and enhance a task.*** Building on (3), teachers can analyze a task for opportunities to develop concepts, procedures, and mathematical processes (among other features of tasks). Such analysis can lead to a discussion of "quick fixes" that increases attention to a Mathematical Practice and conceptual or procedural knowledge. Not all tasks can be "fixed" quickly, but developing a collection of quick ways can help teachers implement the ideas, given the lack of planning time available in day-to-day teaching (see Tool 3.5 or 3.6).

Coaching Lessons From the Field

I was working with a group of emerging coaches related to implementing worthwhile tasks with their teachers. The coaches each selected what they felt was a worthwhile task. Next, they selected one of the planning tools related to worthwhile tasks (see Tools 3.1, 3.2, 3.5, and 3.6). Then, I explained that they were to imagine they were a coachee/teacher and they had received this tool from their coach and were asked to complete it before meeting for a planning conversation so that it could be used as a talking guide. This activity helped these future coaches think about the importance of providing "think time" for teachers.

—Robyn Ovrick
University of Georgia

Connecting to the Leading for Mathematical Proficiency (LMP) Framework

As teachers focus on content knowledge and worthwhile tasks, it is important for them to make explicit connections to the *Shifts in Classroom Practice* and the Mathematical Practices. The brief paragraphs that follow provide ideas for making these connections. Tool 3.1 can also be used for connecting to the *Shifts*.

Connecting Content and Worthwhile Tasks to Shifts in Classroom Practice

Shifts	1	2	3	4	5	6	7	8

Shift 2: From *routine tasks* toward *reasoning tasks*

Teacher uses tasks involving recall of previously learned facts, rules, or definitions and provides students with specific strategies to follow. ⟶ Teacher uses tasks that lend themselves to multiple representations, strategies, or pathways encouraging student explanation (how) and justification (why/when) of solution strategies.

Having a strong understanding of mathematics enables a teacher to "see" the mathematics in a task and consider the possible ways in which it could be represented and solved. Clearly, this impacts their abilities to select and to implement tasks effectively. Content knowledge and task selection go hand in hand: Considering the quality of a task can enhance teacher content knowledge, and studying content more deeply can similarly help teachers select better tasks (see Tools 3.2, 3.5, 3.6, 3.7, and 3.10).

Shift 3: From *teaching about representations* toward *teaching through representations*

Teacher shows students how to create a representation (e.g., a graph or picture). ⟶ Teacher uses lesson goals to determine whether to highlight particular representations or to have students select a representation; in both cases, teacher provides opportunities for students to compare different representations and how they connect to key mathematical concepts.

Representations are not a learning goal in and of themselves; they serve to make mathematics more visible and to help students come to understand such things as why an algorithm works. Consider, for example, how partitioning a rectangle can help students visualize multiplication for whole numbers, decimals, fractions, and algebraic expressions. Here again, we see the importance of teachers understanding such representations and concepts themselves so that they can find tasks with potential and then navigate classroom discussions that connect representations to concepts and procedures (see Tool 3.4).

Shift 6: From *teaching so that students replicate procedures* toward *teaching so that students select efficient strategies*

Teacher approaches facts and procedures with the goal of speed and accuracy. ⟶ Teacher provides time for students to engage with mathematical problems, developing flexibility by encouraging student selection and use of efficient strategies; teacher provides opportunities for students to evaluate when a strategy is best suited for the problem at hand.

This *Shift* was a primary focus in the Coach's Digest. The importance of focusing on developing flexibility cannot be overstated! This does *not* mean that accuracy and speed are not important (this message can be problematic to teachers, administrators, and parents, and it is misguided). But it does mean that focusing solely on accuracy and speed doesn't come close to developing fluency, which must be our goal for every student (see Tools 3.3, 3.7, and 3.9).

Shift 7: From *mathematics-made-easy* toward *mathematics-takes-time*

Teacher presents mathematics in small chunks so that students reach solutions quickly.	Teacher questions, encourages, provides time, and explicitly states the value of grappling with mathematical tasks, making multiple attempts, and learning from mistakes.

"Procedural fluency takes time" might sound like an oxymoron, but it is a true, research-supported reality. Students need time to compare strategies and consider when a particular strategy is a more efficient approach. Teachers need to provide time for students to discuss such ideas before, during, and after they have solved problems.

You can engage teachers in these *Shifts* by focusing specifically on content knowledge or on worthwhile tasks (or both together; see Tool 3.1). There are lots of possibilities for this Focus Zone! For example, with a focus on content knowledge, you might select *Shift* 3, ask teachers to consider the representations for an upcoming unit or lesson, and then consider what teaching moves look like across the continuum. Additionally, or instead, you could focus the discussion on fluency using both *Shifts 3* and *6*, selecting a procedural topic that is on the horizon, and ask teachers to think of things that they might say or do that fall across the continua. To focus on worthwhile tasks, there are many options. You might have them bring a copy of a student page they plan to use, cut up the page into tasks, and place each task on the continuum in terms of its potential to focus on reasoning or to support learning through representations. Or the tasks can be explored in terms of four considerations: what representations are possible, what strategies are possible, what high-level questions fit the task, and to what extent might students grapple with the problem (use a piece of paper folded into fourths to record ideas). When teachers struggle to answer these four questions for a task or worksheet, it can open up a rich conversation about the quality of that task or worksheet. Then, discussion can focus on how to adapt the task(s) or whether to replace them entirely.

Connecting Content and Worthwhile Tasks to Mathematical Practices

MPs 1 2 3 4 5 6 7 8

It is easy to argue that in developing content knowledge, all the Mathematical Practices are relevant, and, in fact, this is certainly true. Here, we describe connections for just four and how they connect to content knowledge and worthwhile tasks.

1. *Make sense of problems and persevere in solving them.* Students must have strong conceptual and procedural knowledge to be able to consider different solution pathways. Additionally, students must have frequent opportunities to engage in higher-level thinking related to all mathematical topics that they learn. As illustrated in Figure 3.2, this will not happen if they are simply told, "This is the procedure for this type of problem—do this."

2. ***Reason abstractly and quantitatively***. Oftentimes, the approach to a problem is based on the numbers in the problem. The answer to the question "How would you solve that proportion problem?" should be "It depends," not "Cross-multiply." Consider these two examples: Which is the better price—4 kiwis for $5.00 or 10 kiwis for $10? Which is the better price—6 oranges for $1.50 or 10 oranges for $3.00? Did you solve them the same way, or did you look at the quantities and take advantage of how they related to other numbers? Such examples can help teachers see the importance of careful and strategic task selection (note that these two examples each lend themselves to a different strategy).

4. ***Model with mathematics***. In order to come up with an equation to represent a situation, students must have a strong understanding of the connection between situations and ways to represent that situation using mathematical representations, such as equations. Making the connection between concrete contexts and abstract mathematics is supported with both conceptual and procedural knowledge, and being asked to make such connections strengthens both kinds of knowledge.

8. ***Look for and express regularity in repeated reasoning***. A major challenge in showing students a way to solve a problem (a standard algorithm or procedure) is that students stop looking for their own methods, shortcuts, patterns, or generalizations. Students (and teachers) may view the new algorithm as *replacing* rather than *adding to* their informal strategies. Yet fluent students select from among strategies, opting for the most efficient strategy. Looking for patterns across tasks can help students (and teachers) focus on *when* a strategy is a good choice, bringing attention to the neglected and critical elements of fluency—strategy selection and flexibility (see Tool 3.3).

→ Ask your teachers to connect the Mathematical Practices to content and/or worthwhile tasks by asking these questions:

- What connections do you see between procedural fluency (all four components) and the selected Mathematical Practices?

- In what ways do the tasks you select influence the opportunities for students to engage in these Mathematical Practices?

- How might you adapt a task or set of problems so that students are able to exhibit these Mathematical Practices? And how do such task adaptations impact procedural fluency?

Coaching Questions for Discussion

Questions Related to the Focus Zone: Content Knowledge and Worthwhile Tasks

1. As you focus on a selected content standard, what might you do to connect it to other content the students have learned or will be learning?

2. Because we have likely learned from and taught from curricula that overemphasize procedures without connections, we need to be intentional about how we approach procedural fluency. What strategies might you use to determine the related concepts and procedures and ensure that you have a good balance of the two?

3. If you are focusing on developing procedural fluency, what might the student and teacher behaviors look like? What might the sample tasks look like? What might the student discourse sound like? How will you connect it to their conceptual knowledge?

4. What does it mean to have a task that is cognitively demanding? What features will it have or not have? What adaptations can you make to a task (in general) that can raise or lower the level of cognitive demand?

5. In considering the objectives of a particular lesson, in what ways do the lesson objectives include conceptual and procedural knowledge and making connections between them? What level of cognition (level of thinking) is expected?

6. In considering the tasks selected for a particular lesson, to what extent does the task develop conceptual and/or procedural knowledge? To what extent does the task engage students in developing fluency? How might the task be adapted to raise the level of cognition?

7. When you introduce the task/lesson, what could you do to ensure that students understand the task and make certain that the level of cognitive demand remains high?

8. In what ways can questions be used to keep the cognitive demand of a task high? How could they (unintentionally) lower the cognitive demand?

9. When students struggle with a task, what actions might you take to help them while retaining the task's high level of cognitive demand? What actions might lower the level of cognitive demand?

10. As you differentiate instruction for your students, how will you ensure that learning focuses on fluency and that the level of thinking remains high?

Questions Related to the LMP Framework

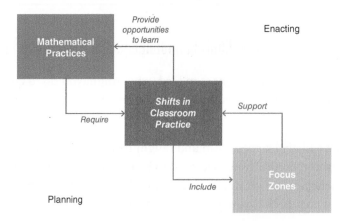

1. As you enact the ideas of content knowledge and worthwhile tasks, what do you see as related *Shifts in Classroom Practice*?

2. As you enact the ideas of content knowledge and worthwhile tasks, what do you see as opportunities to learn and, in particular, to develop the Mathematical Practices?

NOTES

_____
_____
_____
_____
_____
_____
_____
_____
_____
_____
_____
_____
_____
_____
_____
_____
_____
_____
_____

Where to Learn More

Books

NCTM Essential Understanding Series and *Putting Essential Understandings Into Practice* Series (www. nctm.org).

> *These are a series of short books that address a particular topic (e.g., ratios, proportions, and proportional reasoning), identifying the big ideas, connections of the idea to other mathematics, and instructional recommendations. As the titles imply, the first series uses examples and illustrations about the mathematics, and the second series focuses more on implementing the ideas in classrooms.*

National Council of Teachers of Mathematics. (2012). *Rich and Engaging Mathematical Tasks: Grades 5–9*. Reston, VA: NCTM.

> *This book contains a collection of articles and worthwhile mathematical tasks for teachers to use with their students to promote the understanding of the mathematical content highlighted in the Common Core State Standards for Mathematics.*

Schrock, C., Norris, K., Pugalee, D., Seitz R., & Hollingshead, F. (2013). *Great Tasks for Mathematics K–5 and 6–12: Engaging Activities for Effective Instruction and Assessment That Integrate Content and Practices of the CCSS-M*. Aurora, CO: NCSM.

> *The strong collection of tasks in these two books focus on important content and are designed to promote mathematical reasoning. The tasks have also been implemented in classrooms.*

Schuster, L., & Anderson, N.C. (2005). *Good Questions for Math Teaching: Why Ask Them and What to Ask (5–8)*. Sausalito, CA: Math Solutions.

Sullivan, P., & Lilburn, P. (2002). *Good Questions for Math Teaching: Why Ask Them and What to Ask (K–6)*. Sausalito, CA: Math Solutions.

> *While the title of these books says "good questions," it really refers to "good tasks"—and lots of them! You can find many activities that would be great opening tasks for a workshop setting, or particular tasks from the book could be identified for use in an upcoming lesson and could be the focus of a planning session.*

Articles

Drake, C., Land, T. J., Bartell, T. G., Aguirre, J. M., Foote, M. Q., McDuffie, A. R., & Turner, E. E. (2015). "Three Strategies for Opening Curriculum Spaces." *Teaching Children Mathematics, 21*(6), 346–353.

> *Though it uses elementary curriculum as an example, this article can be used at all levels. The authors share how to connect students' mathematical knowledge bases and increase the meaning-making in a lesson by rearranging lesson components, adapting tasks, and making authentic connections.*

Lange, K. E., Booth, J. L., & Newton, K. J. (2014). "Learning Algebra From Worked Examples." *Mathematics Teacher, 107*(7), 534–540.

McGinn, K. M., Lange, K. E., & Booth, J. L. (2015). "A Worked Example for Creating Worked Examples." *Mathematics Teaching in the Middle School, 21*, 26–33.

Using worked examples can improve student achievement and focus on higher-level thinking (evaluation). These two articles provide excellent guidance on using worked examples.

Nagle, C. R., & Styers, J. L. (2015). "Putting Mathematical Tasks Into Context." *Mathematics Teacher, 209*(3), 206–213.

This article shares a professional development activity with high school teachers that contrasts two tasks, one in context and one not in context, to consider the learning opportunities when reasoning contextually. You might do this activity with teachers to support their content development, to discuss the ways in which contexts can build both conceptual and procedural knowledge, and to consider how they might implement the task (or similar one) with students.

NCTM (2014). *Procedural Fluency in Mathematics: A Position of the National Council of Teachers of Mathematics* [Position statement]. Reston, VA: NCTM.

This position paper is a quick read and serves to help understand the full meaning of procedural fluency and how to develop it. A free download from NCTM.

Online Resources

Progressions Documents for the Common Core Math Standards (Institute for Mathematics and Education)

http://ime.math.arizona.edu/progressions

The CCSS for mathematics were built on progressions: narrative documents describing the progression and trajectory of a topic across a number of grade levels. The progressions detail why standards are sequenced the way they are, point out cognitive difficulties, and provide pedagogical solutions and details related to difficult areas of mathematics.

Coach's Toolkit

These tools are a menu from which you can select any that make sense for your setting/context. They can be used independently or as part of a coaching cycle. You may start with the self-assessment, which can guide you in deciding which of the other tools may be most useful. If using these tools for a coaching cycle, mix and match as you like, or use one of the combinations we suggest in the diagrams that follow. The tools include instructions to the coach and the teacher.

Self-Assess

3.1	Connecting *Shifts* to Content and Worthwhile Tasks Self-Assessment

Gather Data

3.7	Developing Mathematical Proficiency
3.8	Implementing Cognitively Demanding Tasks

Plan

3.2	Connecting *Shifts* to Content and Tasks
3.3	Focus on Fluency
3.4	P.I.C.S. Page
3.5	Analyzing Level of Cognitive Demand
3.6	Worthwhile Task Analysis

Reflect

3.9	Impact on Students' Emerging Fluency
3.10	Reflecting on Task Implementation

Topic:
Worthwhile Tasks

Plan
**3.2 or 3.5
or 3.6**

Reflect
3.10

Gather
Data
3.8

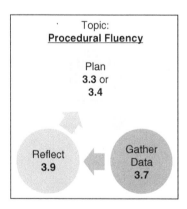

Topic:
Procedural Fluency

Plan
**3.3 or
3.4**

Reflect
3.9

Gather
Data
3.7

Additional Tools in Other Chapters

Implementing high-quality tasks is also the focus of Chapter 5 (Questioning and Discourse; see Tools 5.3, 5.4, and 5.5), Chapter 7 (Analyzing Student Thinking), and Chapter 8 (Differentiating Instruction, which includes differentiating tasks).

 To download the coaching tools for Chapter 3 that only have instructions for the teacher, go to **resources.corwin.com/mathematicscoaching**.

3.1 Connecting *Shifts* to Content and Worthwhile Tasks Self-Assessment

Instructions to the Coach: Ask teachers (individually or as part of a PLC activity) to self-assess where they position themselves on each of these Shifts in Classroom Practice *related to content and worthwhile tasks. Use the questions that follow during a coaching conversation or in a PD setting to support teachers (and to help you decide which tools may be most useful from this chapter). A one-page version of this tool without this note is available for download.*

Instructions: The *Shifts in Classroom Practice* that follow have specific connections to content knowledge and worthwhile tasks. Put an *X* on the continuum of each *Shift* to identify where you currently see your practice.

Tool 3.1 Shifts

Shift 2: From *routine tasks* toward *reasoning tasks*

Teacher uses tasks involving recall of previously learned facts, rules, or definitions and provides students with specific strategies to follow.	Teacher uses tasks that lend themselves to multiple representations, strategies, or pathways encouraging student explanation (how) and justification (why/when) of solution strategies.

Shift 3: From *teaching about representations* toward *teaching through representations*

Teacher shows students how to create a representation (e.g., a graph or picture).	Teacher uses lesson goals to determine whether to highlight particular representations or to have students select a representation; in both cases, teacher provides opportunities for students to compare different representations and how they connect to key mathematical concepts.

Shift 6: From *teaching so that students replicate procedures* toward *teaching so that students select efficient strategies*

Teacher approaches facts and procedures with the goal of speed and accuracy.	Teacher provides time for students to engage with mathematical problems, developing flexibility by encouraging student selection and use of efficient strategies; teacher provides opportunities for students to evaluate when a strategy is best suited for the problem at hand.

Shift 7: From *mathematics-made-easy* toward *mathematics-takes-time*

Teacher presents mathematics in small chunks so that students reach solutions quickly.	Teacher questions, encourages, provides time, and explicitly states the value of grappling with mathematical tasks, making multiple attempts, and learning from mistakes.

Tool 3.1 Reflection Questions

1. What do you notice, in general, about your self-assessment of these *Shifts in Classroom Practice*?

2. What might be specific teaching moves that align with where you placed yourself on the *Shifts*?

3. What might be specific teaching moves that align *to the right of* where you placed yourself on the *Shifts*?

4. What might be some professional learning opportunities to help you move to the right for one or more of these *Shifts*?

3.2 Connecting *Shifts* to Content and Tasks

Instructions to the Coach: As an alternative or as a follow-up to Tool 3.1, use this tool to help teachers see how the Shifts apply to a particular lesson. In a coaching cycle, teachers could complete this in preparation for a planning conversation, or they could just bring their task with them and the planning conversation can be the questions provided in the template.

Instructions: The *Shifts in Classroom Practice* can provide support in keeping the level of cognitive demand high through task implementation. Select one or two *Shifts* as a focus and then complete the planning table here.

Shift(s):

Toward . . . _____

Toward . . . _____

The Task	Classroom Environment	Setting Up the Task
How might you adapt the task to reflect the selected *Shifts*?	How might you organize students?	How might you pose the task to reflect the selected *Shift(s)*?
Lesson	**Discussing the Task**	**Assessing**
How might you structure the lesson itself to align with the *Shift(s)*?	What questions or questioning strategies will you use?	What will you be looking for as students work?

Source: *Previously published by Bay-Williams, J., McGatha, M., Kobett, B., and Wray, J. (2014).* Mathematics Coaching: Resources and Tools for Coaches and Leaders, K–12. *New York, NY: Pearson Education, Inc.*

3.3 Focus on Fluency

Instructions to the Coach: Ask teachers (individually or as part of a PLC activity) to complete this tool for an upcoming procedure-related topic as a form of topic study. Use this completed tool for a planning conversation to consider where elements of fluency will be attended to in a lesson.

Instructions: Analyze a procedure-related topic, identifying how components of fluency will be included (e.g., tasks, questions to pose). Note that appropriate strategy selection is nested *within* efficiency to emphasize that efficiency is *not* only about speed.

Procedure: _____

Efficiency	Flexibility
Speed/automaticity	**Strategy adaptation**
Strategy selection	**Strategy transference**
Accuracy	
Correct solutions	

3.4 P.I.C.S. Page

Instructions to the Coach: The P.I.C.S. Page can be used in several ways: analyzing a unit to identify important content across these four domains; designing a lesson or unit (ensuring adequate attention to each domain); considering purposeful questions in each domain; designing pre- and/or post-assessments for students. Select any of these to use with teachers and ask them how they might use this tool with students.

Instructions: Identify a topic and provide examples and illustrations in each box.

Topic: _____

Procedure (symbolic representation)	Illustration (picture or manipulative)

Concept (general explanation)	Situation (context)

Source: *Previously published by Bay-Williams, J., McGatha, M., Kobett, B., and Wray, J. (2014).* Mathematics Coaching: Resources and Tools for Coaches and Leaders, K–12. *New York, NY: Pearson Education, Inc.*

3.5 Analyzing Level of Cognitive Demand

Instructions: Use the levels of cognitive demand to evaluate a task or lesson. Review the descriptors and highlight those that match the task you have selected.

Low-Level Cognitive Demand
Memorization Tasks
• Involve either memorizing or producing previously learned facts, rules, formulae, or definitions • Are routine, involving exact reproduction of previously learned procedure • Have no connection to related concepts
Procedures Without Connections Tasks
• Specifically call for use of the procedure • Are straightforward, with little ambiguity about what needs to be done and how to do it • Have no connection to related concepts • Focus on producing correct answers, rather than on developing mathematical understanding • Require no explanations, but focus on the procedure only
High-Level Cognitive Demand
Procedures With Connections Tasks
• Focus students' attention on the use of procedures for the purpose of developing deeper levels of understanding of mathematical concepts and ideas • Suggest general procedures that have close connections to underlying conceptual ideas • Are usually represented in multiple ways (e.g., visuals, manipulatives, symbols, problem situations) • Require that students engage with the conceptual ideas that underlie the procedures in order to successfully complete the task
Doing Mathematics Tasks
• Require complex and non-algorithmic thinking (i.e., nonroutine—there is not a predictable, known approach) • Require students to explore and to understand the nature of mathematical concepts, processes, or relationships • Demand self-monitoring or self-regulation of cognitive processes • Require students to access relevant knowledge in working through the task • Require students to analyze the task and actively examine task constraints that may limit possible solution strategies and solutions • Require considerable cognitive effort

Source: *Adapted from Smith, M. S., and Stein, M. K. (1998). "Selecting and Creating Mathematical Tasks: From Research to Practice."* Mathematics Teaching in the Middle School, 3(5): 344–350. Previously published by Bay-Williams, J., McGatha, M., Kobett, B. & Wray, J. (2014) Mathematics Coaching: Resources and Tools for Coaches and Leaders, K–12. New York, NY: Pearson Education, Inc.

1. Describe your overall evaluation of whether this task/lesson has the potential to engage students in higher-level thinking.

2. What adaptations can you make to the task or lesson to increase its higher-level thinking potential?

Source: Previously published by Bay-Williams, J., McGatha, M., Kobett, B., and Wray, J. (2014). Mathematics Coaching: Resources and Tools for Coaches and Leaders, K–12. New York, NY: Pearson Education, Inc.

3.6 Worthwhile Task Analysis

Instructions to the Coach: This tool is very effective in professional learning to focus on analyzing and adapting tasks. And it can be used in a coaching cycle.

Instructions: Using the Worthwhile Tasks Focus Zone prompts below (NCTM, 1991; NCTM, 2007), rate a task you are planning to use in a lesson. Add comments about how it might be adapted to better address the stated quality of a worthwhile task.

1 = No evidence of the quality in the task, or it is not possible to address this quality with this task.

2 = The quality is evident in minor ways, or incorporating it is possible.

3 = The quality is evident in the task.

4 = The quality is central to the task and is important to the success of the lesson.

Aspects of a Worthwhile Task	Rating				How I Might Enhance Task
Mathematics in the task is powerful.					
1. Is grade or course-level appropriate	1	2	3	4	
2. Makes connections between concepts and procedures (high cognitive level)	1	2	3	4	
3. Makes connections between different mathematical topics	1	2	3	4	
4. Requires reasoning (non-algorithmic thinking)	1	2	3	4	
Task is connected to the student.					
5. Connects to real situations that are familiar and relevant to them	1	2	3	4	
6. Provides multiple entry points that make it accessible to each student	1	2	3	4	
7. Is appropriately challenging (engages students' interests and intellect)	1	2	3	4	
Task lends to observing and assessing student understanding.					
8. Provides multiple ways to demonstrate understanding of the mathematics	1	2	3	4	
9. Requires students to illustrate or explain mathematical ideas	1	2	3	4	
10. Has potential to develop perseverance and positive student dispositions	1	2	3	4	

1. Describe your overall evaluation of whether this task/lesson has the potential to engage students in higher-level thinking.

2. What adaptations can you make to the task or lesson to increase its higher-level thinking potential?

3.7 Developing Mathematical Proficiency

*Instructions to the Coach: These strands continue to be a useful tool for thinking about content and mathematical practices. Share the table during a planning conversation. Record data during the lesson using the chart or using sticky notes. In a reflecting conversation, discuss how the student moves connect to the **five strands** and what next steps might increase the focus on some strands.*

Instructions: Record student actions, including verbal and written. After the lesson, consider how these actions map to the five strands.

Five Strands of Mathematical Proficiency (NRC, 2001)

1. Conceptual Understanding	To what extent do students understand the concepts? Do students understand the operations and procedures they are using? Do they see relationships or connections between ideas?
2. Procedural Proficiency	To what extent do students use procedures flexibly and/or efficiently? Do students select a computational approach that best suits the problem (efficient method) or solve all problems of the same type the same way?
3. Strategic Competence	In what ways do students demonstrate that they can come up with their own approach to a problem? Do students choose and use representations to support their thinking? Do they pick appropriate approaches to solve the problem?
4. Adaptive Reasoning	In what ways do students monitor their problem-solving, seeing it if is working for them and changing the process as needed to reach a solution? Do students abandon one approach and pursue another?
5. Productive Disposition	To what extent do students demonstrate a "habit of mind" that mathematics makes sense and is useful? Do students demonstrate confidence in designing their own solution strategies and persevering as they encounter challenging problems?

Data Gathering

Time	Student Statement, Action, or Observed Work	Mathematical Proficiency (check all that apply)
		☐ Conceptual Understanding ☐ Procedural Proficiency ☐ Strategic Competence ☐ Productive Disposition ☐ Adaptive Reasoning
		☐ Conceptual Understanding ☐ Procedural Proficiency ☐ Strategic Competence ☐ Productive Disposition ☐ Adaptive Reasoning
		☐ Conceptual Understanding ☐ Procedural Proficiency ☐ Strategic Competence ☐ Productive Disposition ☐ Adaptive Reasoning
		☐ Conceptual Understanding ☐ Procedural Proficiency ☐ Strategic Competence ☐ Productive Disposition ☐ Adaptive Reasoning
		☐ Conceptual Understanding ☐ Procedural Proficiency ☐ Strategic Competence ☐ Productive Disposition ☐ Adaptive Reasoning

3.8 Implementing Cognitively Demanding Tasks

Instructions to the Coach: During the lesson, record tasks used and how they were implemented. After the lesson, reflect on the level of cognitive demand of the task itself and the cognitive demand in its implementation.

Task(s) Selection (record or describe)	Task(s) Implementation (record teacher actions, student actions)	Cognitive Demand (discuss in reflecting conversation)

Codes for Type of Task:
Memorization = M
Procedures With Connections = PWC
Procedures With No Connections = PNC
Doing Mathematics = DM

3.9 Impact on Students' Emerging Fluency

Instructions to the Coach: This tool can be used as a follow-up to Tool 3.7. Teachers can have it already filled out for a reflecting conversation, or you can use it as a discussion prompt.

Instructions: Discuss or write responses to these prompts as a way to reflect on procedural fluency, including student learning of related concepts and procedures (and relationships between them).

1. Based on the data gathered, in what ways did *you* focus on each element of fluency?
   Flexibility:

   Strategy Selection:

   Efficiency:

   Accuracy:

2. Based on the data gathered, in what ways did *students* demonstrate each element of fluency?
   Flexibility:

   Strategy Selection:

   Efficiency:

   Accuracy:

3. In thinking about the lesson just taught, what teacher moves were particularly effective in eliciting student actions that demonstrated mathematical proficiency?

4. Which of these moves or new moves might be used in future lessons to support procedural fluency?

3.10 Reflecting on Task Implementation

Instructions to the Coach: This tool can be used as a follow-up to Tool 3.8. If not connected with Tool 3.8, begin by having the teacher review and sort teacher actions and student actions from the lesson.

Instructions: Discuss the following, using evidence and examples from the data collected.

1. In what ways did you feel that the implementation of the task resulted in its being a "worthwhile" opportunity for students to learn important mathematics?

2. How effective was the implementation of the task in engaging students with the context of the problem?

3. How effective was the implementation of the task in helping students understand concepts, procedures, and connections between the two?

4. To what extent was each student appropriately challenged and engaged in productive struggle?

5. In looking at the data, what teacher moves did you make that may have maintained a high level of cognitive demand and/or reduced the high level of cognitive demand? What was effective?

6. What general insights about Effective Mathematics Teaching did this lesson cycle provide?

Source: *Previously published by Bay-Williams, J., McGatha, M., Kobett, B., and Wray, J. (2014).* Mathematics Coaching: Resources and Tools for Coaches and Leaders, K–12. *New York, NY: Pearson Education, Inc.*

Chapter 4

Engaging Students

Dear Coach,

Here is support for you as you work with teachers on engaging students.

In the **COACH'S DIGEST**...

Overview: A review of brain research and how it can influence the engagement techniques used with students. Also available as a download to share with teachers.

Coaching Considerations for Professional Learning: Ideas for how to support a teacher or a group of teachers in learning about engagement techniques to support student learning.

Coaching Lessons From the Field: Mathematics coaches share how they used a tool or idea from this chapter.

Connecting to the Framework: Specific ways to connect selected *Shifts* and Mathematical Practices to student engagement.

Coaching Questions for Discussion: Menu of prompts about student engagement for professional learning or one-on-one coaching.

Where to Learn More: Articles, books, and online resources for you and your teachers on where to learn more about engaging students!

In the **COACH'S TOOLKIT**...

Eight tools focused on engagement techniques for professional learning or coaching cycles.

Coach's Digest

In the Coach's Digest, we begin with an overview of engaging students, written to teachers (and to you, the coach). As you read the Overview, the following questions might help you reflect on this topic in terms of your role as a coach:

- If I build on my teachers' strengths, as I am asking them to do with their students, where might I start in implementing these strategies?
- What types of professional learning activities, readings, and/or coaching cycles and lesson study might support their efforts to implement these ideas?
- What data might we collect to see if the strategies are indeed engaging students?

To download the Chapter 4 Overview, go to **resources.corwin.com/mathematicscoaching.**

What does *engaging students* really mean? What do terms such as ***active engagement, total participation techniques, thinking strategies, student engagement, cooperative learning,*** and ***whole-class discussions*** all have in common? In general, all of these ideas are focused on moving away from students as passive learners. Advances in brain imaging have provided new insights into how students learn, which has prompted new discussions about teaching. Jensen (2008) describes brain-based learning in three words: "engagement, strategies, and principles" (p. 4). In other words, it is the engagement of strategies based on [principles of] how the brain works. While a variety of brain research has provided many insights relative to teaching, we will focus on two "big ideas" here that relate specifically to engaging students: (1) Physical activity is critical to learning, and (2) learning is a social endeavor.

Physical and Social Engagement

Physical activity elevates the brain chemicals that affect thinking and learning (Erikson, Hillman, & Kramer, 2015; Jensen, 2008). This is one reason many schools are reinstituting recess (Strauss, 2016). While it may not be possible to change recess or school schedules, it is possible to add physical movement to mathematics teaching. One easy-to-implement strategy is the use of **learning partners**—students find their own or prearranged partner and meet with him or her to discuss ideas about a task. By asking students to move and find a place to have a stand-up conversation (instead of turn and talk to their neighbor), physical movement has been incorporated. Of course, this is not the same effect as 30 minutes on the playground, but it can impact students' motivation and learning.

Teachers have known for some time that working in cooperative groups and using whole-class discussions are valuable to support students' learning. Through recent brain research, we know that positive social classroom activities can positively influence the brain; however, negative classroom social interactions and relationships can have the reverse effect (Jensen, 2008; Lieberman, 2014). Jensen (2008) recommends using targeted, planned small groups (as in cooperative learning) in which positive relationships can be fostered.

Himmele and Himmele (2017) combine a focus on brain research (participation) and high-level thinking (cognition) in their **Total Participation Techniques (TPT) Cognitive Engagement Model** shown in Figure 4.1.

The intersection of the two continua produces four quadrants. Quadrant 1 teaching is low-level thinking with little participation from students—for example, students passively listening (at least some of them) to their teacher telling them how to solve a problem. Mathematics teaching can get stuck in Quadrant 1, severely limiting student learning. Quadrant 2 teaching actively engages students, but the tasks are at low cognitive levels. Many "fun" math activities actually fall into this quadrant. At face value, it may seem as if students are engaged, but it is not likely their brain is engaged in mathematical reasoning. Quadrant 3 teaching is using high-level thinking tasks or questions but only engaging a handful of students. Quadrant 4 teaching engages all students and focuses on high-level thinking. It is important to point out that aspects of teaching could fall within all four quadrants, but it is critical to spend significant time in Quadrant 4 to ensure *all* students are learning at high levels.

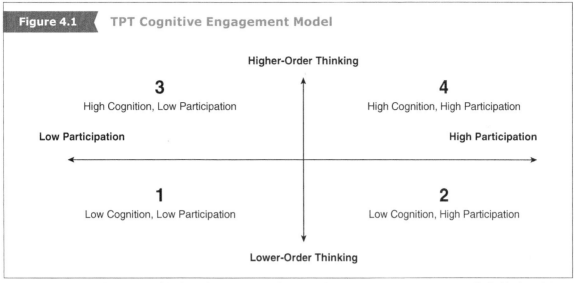

| Figure 4.1 | TPT Cognitive Engagement Model |

Higher-Order Thinking

3
High Cognition, Low Participation

4
High Cognition, High Participation

Low Participation

High Participation

1
Low Cognition, Low Participation

2
Low Cognition, High Participation

Lower-Order Thinking

Source: Himmele, P., and Himmele, W. (2017). Total Participation Techniques: Making Every Student an Active Learner *(2nd ed.). Alexandria, VA: ASCD.*

Engagement Techniques

The rise in attention to brain-based/engaged learning has led to many techniques that teachers can infuse into their teaching to support high participation in their students. Himmele and Himmele (2017) suggest 51 total participation techniques, such as one-word summaries, quick-writes, and bounce cards. They include suggestions on how to ensure high-level thinking with each technique. Another source is the Thinking Collaborative, which includes over 160 techniques for engagement as well as facilitator moves (www.thinkingcollaborative.com). In fact, there is a nearly endless list on the Internet. It is important to remember to combine these **engagement techniques** with high-level thinking to optimize student learning.

NOTES

_____

_____

_____

_____

_____

_____

_____

_____

_____

_____

_____

_____

_____

_____

_____

_____

_____

Coaching Considerations for Professional Learning

Simply having a list of effective engagement techniques is not enough! Teachers need support in understanding the intent of the engagement techniques and integrating the techniques into their practice. For some teachers, these engagement techniques will represent a paradigm shift in what it means to teach mathematics. What follows are some ideas for supporting teachers in the process.

1. ***Engagement techniques self-assessment.*** Using one of the engagement techniques resources discussed in this chapter (e.g., Thinking Collaborative, Total Participation Techniques), ask teachers to self-assess their use of selected techniques on a scale such as novice–apprentice–proficient–distinguished. The self-assessment can be used to plan professional development to support teachers in growth areas. Alternatively, have teachers share engagement techniques they use, create a master list, and use it for the self-assessment. See Tool 4.2 to help teachers plan for engagement techniques using the TPT Cognitive Engagement Model.

2. ***Engagement techniques in action.*** It helps teachers to see student engagement (with high-level thinking) in practice. View videos of teaching and ask teachers to reflect on (1) techniques implemented in the lesson and (2) how effective the techniques were in student participation and high-level thinking. Some video resources include the following:
   - PRIMAS (http://primas.mathshell.org)
   - NCTM Illuminations (www.illuminations.nctm.org)
   - Teaching Math: A Video Library (www.learner.org)
   - Inside Mathematics (www.insidemathematics.org)
   - Edutopia (www.edutopia.org/blogs/tag/student-engagement)

   When teachers are comfortable with integrating engagement techniques into their practice, make your own collection of videos to share in your district.

3. ***Engagement techniques sort.*** This activity can be conducted in a PLC setting. Decide on a certain number of days to collect lessons/activities (the minimum should be one week). Ask each teacher to bring in sample lessons or instructional activities from each day they taught within the designated time period. Ask each teacher to share and then have the group sort the activities. Leaving the directions open-ended and allowing teachers to decide on sorting criteria can lead to rich discussions. Ask teachers to describe how the data from the sorting activity (number of activities in each category, categories, etc.) might inform their practice. Alternatively, sort the activities using the TPT Cognitive Engagement Model (see Tool 4.5).

4. ***Cooperative learning.*** Cooperative learning has been used by teachers for many years. However, it is still a relevant engagement technique that focuses on the second big idea shared in the Overview ("Learning is a social endeavor"). Use Tools 4.3, 4.5, 4.7, and 4.8 to support teachers in effectively implementing cooperative learning.

Coaching Lessons From the Field

When presenting a professional development session, I typically model and incorporate a number of engagement strategies, which are recorded on anchor charts for reference. Toward the end of the session, I then ask the teachers to partner and conduct a "strategy harvest" (www.thinkingcollaborative.com). The strategy harvest allows session participants the opportunity to review the anchor chart lists, clarify the different strategies and the purpose for which each was used within the session, and share possible applications or adaptations for use with students in the classroom. During a whole-group debrief, I answer any clarifying questions about the strategies and ask teachers to share how they have used, or might use, different ones to engage students and/or foster deeper thinking about the content. Teachers not only appreciate the opportunity to experience and examine the strategies from a learner stance, but they also leave with a list of new "tools" they can implement in their classrooms.

—Candy Thomas,
Instructional Coach Oldham County, KY

Connecting to the Leading for Mathematics Proficiency (LMP) Framework

As teachers focus on engaging students, it is important for them to make explicit connections to the *Shifts in Classroom Practice* and the Mathematical Practices. The following brief paragraphs provide ideas for making these connections. You and your teachers can continue to add to these ideas (see also Tool 4.1 for connecting to the *Shifts*).

Connecting Engaging Students to *Shifts in Classroom Practice*

Shifts   1   **2**   3   **4**   **5**   6   **7**   **8**

Shift 2: From *routine tasks* toward *reasoning tasks*

Teacher uses tasks involving recall of previously learned facts, rules, or definitions and provides students with specific strategies to follow.	$\longrightarrow$	Teacher uses tasks that lend themselves to multiple representations, strategies, or pathways encouraging student explanation (how) and justification (why/when) of solution strategies.

Reasoning tasks provide opportunities for students to engage in making sense of mathematics. These high-level tasks lend themselves to a variety of engagement techniques that encourage students to analyze and critique various solution strategies (see Tool 4.2). And by using engagement techniques, students have more of an opportunity to explain and justify their solution strategies.

Shift 4: From *show-and-tell* toward *share-and-compare*

Teacher has students share their answers.	$\longrightarrow$	Teacher creates a dynamic forum where students share, listen, honor, and critique each other's ideas to clarify and deepen mathematical understandings and language; teacher strategically invites participation in ways that facilitate mathematical connections.

This *Shift* embodies everything about engaging the student! On the left end is Quadrant 1 (Figure 4.1), low-engagement, low-level thinking. On the right end is Quadrant 4, a "dynamic forum" where students are engaged in mathematics reasoning and learning. Teachers must be intentional in inviting this kind of participation; it does not happen without planning (see Tools 4.2 and 4.3).

Shift 5: From *questions that seek expected answers* toward *questions that illuminate and deepen student understanding*

Teacher poses closed and/or low-level questions, confirms correctness of responses, and provides little or no opportunity for students to explain their thinking.	Teacher poses questions that advance student thinking, deepen students' understanding, make the mathematics more visible, provide insights into student reasoning, and promote meaningful reflection.

In the TPT Cognitive Engagement Model (Figure 4.1), one-half of the criteria focus on high-level thinking—this includes posing questions that illuminate and deepen thinking. Without the focus on high-level questions, the engagement can move to Quadrant 2, "fun" activities but without the cognitive rigor necessary to move students forward in their thinking (see Tools 4.2 and 4.4). Additionally, teachers can ask fantastic questions that have the potential to deepen students' understanding, but not make sure every student is thinking of the answer. Engagement techniques can fix this problem, moving instruction from Quadrant 3 to Quadrant 4.

Shift 7: From *mathematics-made-easy* toward *mathematics-takes-time*

Teacher presents mathematics in small chunks so that students reach solutions quickly.	Teacher questions, encourages, provides time, and explicitly states the value of grappling with mathematical tasks, making multiple attempts, and learning from mistakes.

Productive struggle is all about engaging students in making sense of problem situations. Teachers can use a variety of engagement techniques to support students in productive struggle, such as posing questions, allowing students time to think, and providing time to talk with peers. For example, after students have read the task they are going to solve, a teacher might ask them to go find their learning partner just to talk about the problem (not how to solve it). When they return to their seats, they can exchange what they discussed with an elbow partner.

Shift 8: From *looking at correct answers* toward *looking for students' thinking*

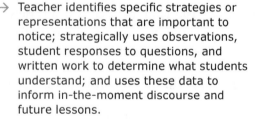

Teacher attends to whether an answer or procedure is (or is not) correct.	Teacher identifies specific strategies or representations that are important to notice; strategically uses observations, student responses to questions, and written work to determine what students understand; and uses these data to inform in-the-moment discourse and future lessons.

Using engagement techniques provides the teacher many creative ways to focus on students' thinking. In fact, as teachers infuse more engagement techniques into their teaching, they will see much more evidence of student thinking. Cooperative learning also provides much more access to student thinking. Teachers can use one of the formative assessment tools described in Chapter 6 as they observe students working in groups.

Engagement is key to learning. Engaged students persevere and engage in productive struggle. Yet a common concern for teachers is unmotivated students. Sometimes, that lack of motivation is due to how the content is presented. Providing concrete strategies for engaging students and asking teachers to practice these strategies and reflect on how to improve the use of these strategies can result in a much more engaged classroom. During professional learning, teachers can share what engagement strategies they used, along with implementation successes and challenges.

Connecting Engaging Students to Mathematical Practices

MPs   1   2   **3**   4   **5**   **6**   7   8

The Mathematical Practices are about what students must be able to do. Certainly, that has strong implications for teachers trying to engage all students. For example, how can students evaluate the reasoning of other students if the teacher has shown everyone how to do a problem, leaving little to be created by students and thus little to be evaluated? The mathematical proficiencies described in the following three Mathematical Practices will only be developed if students are actively engaged and thinking at high levels (Quadrant 4, Figure 4.1) on a daily basis.

3. *Construct viable arguments and critique the reasoning of others.* To construct arguments and critique the reasoning of others, students need opportunities to engage with other students about rich mathematics tasks that encourage multiple approaches. Effective high-level questioning by the teacher supports students in learning to critique their own and others' reasoning. Engagement techniques can be used to provide variety in how students are engaging with each other about their mathematical reasoning and solution strategies. Effective cooperative groups provide an opportunity to vet ideas as they are emerging, as well as to defend strategies to a smaller audience prior to a whole-group discussion.

5. *Use appropriate tools strategically.* When students use physical tools, they are physically active. But simply using physical tools is not enough; the mathematics is not in the tool. Teachers must support students in connecting the physical tool with the mathematics in the task. With carefully selected small-group tasks that use tools, teachers can have students physically engaged, interacting socially, and reasoning at high levels. When students select different tools (or representations), they have something to share (social engagement) and compare (high-level thinking).

6. *Attend to precision.* Attending to precision pertains to both verbal and written communication. In order for students to communicate verbally in precise ways, they must have opportunities to engage in mathematical discussions. Students who are struggling readers or who are learning English need explicit attention to the precise language of mathematics, as well as opportunities to use that language with others. Engagement techniques foster low-risk opportunities to improve students' ability to communicate mathematically.

> → Ask your teachers to connect the Mathematical Practices to engaging students by asking these questions:
>
> - What connections do you see between engagement techniques and the selected Mathematical Practices?
>
> - What adjustments have you made in your engagement techniques that support students in demonstrating the Mathematical Practices?
>
> - How does the engagement during your lessons influence the opportunities for students to demonstrate the Mathematical Practices?

NOTES

_____

_____

_____

_____

_____

_____

_____

_____

_____

_____

_____

_____

_____

_____

_____

_____

_____

_____

_____

_____

_____

_____

_____

_____

Coaching Questions for Discussion

Questions Related to the Focus Zone: Engaging Students

1. Knowing the benefits of actively engaging students, what might be some strategies you could use to increase engagement techniques in your practice?

2. In what ways could you incorporate more physical movement into your teaching?

3. What strategies might you use to be sure all students are engaged in thinking of the answers to the questions you ask?

4. In what ways do you monitor who is engaged in classroom discussions?

5. How might your PLC support one another in implementing engagement techniques?

6. How might you build in opportunities for students to talk about mathematics in your lessons?

Questions Related to the LMP Framework

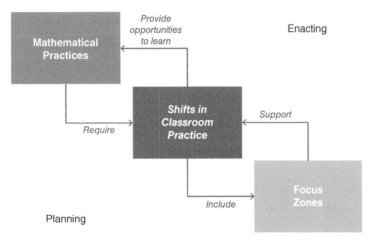

1. As you enact the ideas of engaging students, what do you see as related *Shifts in Classroom Practice?*

2. As you enact the ideas of engaging students, what do you see as opportunities to learn and, in particular, to develop the Mathematical Practices?

Where to Learn More

Books

Himmele, P., & Himmele, W. (2017). *Total Participation Techniques: Making Every Student an Active Learner.* Alexandria, VA: ASCD.

> *This book describes the TPT Cognitive Engagement Model and the brain research upon which the model is based. The authors then introduce 51 engagement techniques. For each, they provide a description, how the technique works, how to maintain the high-order thinking, and suggestions for adapting and personalizing the technique.*

Hoffer, W. W (2012). *Minds on Mathematics: Using Math Workshop to Develop Deep Understanding in Grades 4–8.* Portsmouth, NH: Heinemann.

> *The author describes the benefits of creating a "minds-on" math workshop where students learn mathematics at deep levels. She outlines how teachers serve as facilitators of the math workshop to support students in building a community of learners. In addition, she provides detailed strategies for implementing a math workshop.*

Horn, I. S. (2017). *Motivated: Designing Math Classrooms Where Students Want to Join In.* Portsmouth, NH: Heinemann.

> *This book describes five features of motivational math classrooms: belongingness, meaningfulness, competence, accountability, and autonomy. The author provides suggestions for including each aspect in math instruction in order to motivate student engagement. The reader is introduced to six teachers and their experiences in creating motivational math classrooms.*

Jensen, E. (2008). *Brain-Based Learning: The New Paradigm of Teaching.* Thousand Oaks, CA: Corwin.

> *This book presents an overview of recent brain research and how it should impact the way we teach. The author presents a set of overarching principles based on brain research to guide teachers as they implement strategies that will engage students. Although not specific to mathematics, the suggestions are applicable for mathematics teachers.*

McCoy, A., Barnett, J., & Combs, E. (2013). *High-Yield Routines for Grades K–8.* Reston, VA: NCTM.

> *The authors present seven high-yield routines that can enhance students' engagement in math classrooms. With each routine, they share a vignette that provides an example of the routine in action. This is followed by a description of the routine and strategies for implementation. Examples of student work are included, with suggestions for assessing and adapting the routines.*

Articles

McFeetors, P. J., & Palfy, K. (2017). "We're in Math Class Playing Games, Not Playing Games in Math Class." *Mathematics Teaching in the Middle School, 22*(9), 534–544.

> *The authors describe how they engaged middle school students in playing commercial strategy games to increase students' reasoning skills. They share two explicit pedagogical practices they used to support an environment where students developed reasoning skills in collaborative settings.*

Watanabe, M., & Evans, L. (2015). "Assessments That Promote Collaborative Learning." *Mathematics Teacher, 109*(4), 298–304.

> *This article describes participation quizzes and explanation quizzes used by a high school mathematics department to encourage students to collaborate and take responsibility for one another's learning. The authors provide suggestions for selecting tasks for these quizzes and how to build a collaborative community.*

Zhang, X., Clements, M. A., & Ellerton, N. F. (2015). "Engaging Students With Multiple Models." *Teaching Children Mathematics, 22*(3), 139–147.

> *The authors present three multimodal models with six activities designed to actively engage students in building a conceptual understanding of unit fractions. The activities were used in a research study with fifth graders, and those students who engaged in the multimodal activities developed deeper conceptual understandings of unit fractions.*

Online Resources

Edutopia

www.edutopia.org/blogs/tag/student-engagement

> *This site has several blogs focused on student engagement, as well as videos and additional resources.*

Thinking Collaborative

www.thinkingcolaborative.com

> *This site includes over 160 techniques for engagement, as well as facilitator moves. Bookmark it and use it for planning your professional learning events and to share with teachers.*

NOTES

Coach's Toolkit

These tools are a menu from which you can select any that make sense for your setting/context. They can be used independently or as part of a coaching cycle. You may start with the self-assessment, which can guide you in deciding which of the other tools may be most useful. If using these tools for a coaching cycle, mix and match as you like or use the combinations we suggest in the diagrams that follow. The tools in this chapter include instructions to the coach and the teacher. You can download copies of the tools that only have instructions for the teacher at **resources.corwin.com/ mathematicscoaching.**

Self-Assess

4.1	Connecting *Shifts* to Engaging Students Self-Assessment

Gather Data

4.4	Engagement Techniques
4.5	Cooperative Groups Data Gathering

Plan

4.2	Total Participation Technique (TPT) Planning
4.3	Planning for Cooperative Groups

Reflect

4.6	Engagement Techniques Discussion Prompts
4.7	Reflecting on Cooperative Groups
4.8	Analyzing Learning in Cooperative Groups

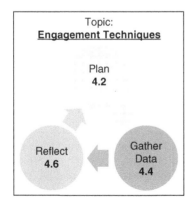

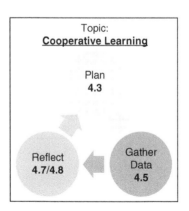

Additional Tools in Other Chapters

Chapter 2 (Effective Mathematics) addresses the *Shifts* and Mathematical Practices broadly—so many of the tools in that chapter address engagement techniques. Chapter 5 (Questioning and Discourse, focusing on asking high-level questions) and Chapter 6 (Formative Assessment, focusing on embedding formative assessment in instruction) also offer tools to support teachers' use of engagement techniques.

online resources 🐾 To download the coaching tools for Chapter 4 that only have instructions for the teacher, go to **resources.corwin.com/mathematicscoaching.**

4.1 Connecting *Shifts* to Engaging Students
Self-Assessment

Instructions to the Coach: Ask teachers (individually or as part of a PLC activity) to self-assess where they position themselves on each of these Shifts in Classroom Practice *related to engaging students. Use the questions that follow during a coaching conversation or in a PD setting to support teachers (and to help you decide which tools may be most useful from this chapter). A one-page version of this tool without this note is available for download.*

Instructions: The *Shifts in Classroom Practice* listed below have specific connections to engaging students. Put an *X* on the continuum of each *Shift* to identify where you currently see your practice.

Tool 4.1 Shifts

Shift 2: From *routine tasks* toward *reasoning tasks*

Teacher uses tasks involving recall of previously learned facts, rules, or definitions and provides students with specific strategies to follow.	Teacher uses tasks that lend themselves to multiple representations, strategies, or pathways encouraging student explanation (how) and justification (why/when) of solution strategies.

Shift 4: From *show-and-tell* toward *share-and-compare*

Teacher has students share their answers.	Teacher creates a dynamic forum where students share, listen, honor, and critique each other's ideas to clarify and deepen mathematical understandings and language; teacher strategically invites participation in ways that facilitate mathematical connections.

Shift 5: From *questions that seek expected answers* toward *questions that illuminate and deepen student understanding*

Teacher poses closed and/or low-level questions, confirms correctness of responses, and provides little or no opportunity for students to explain their thinking.	Teacher poses questions that advance student thinking, deepen students' understanding, make the mathematics more visible, provide insights into student reasoning, and promote meaningful reflection.

Shift 7: From *mathematics-made-easy* toward *mathematics-takes-time*

Teacher presents mathematics in small chunks so that students reach solutions quickly.	Teacher questions, encourages, provides time, and explicitly states the value of grappling with mathematical tasks, making multiple attempts, and learning from mistakes.

Shift 8: From *looking at correct answers* toward *looking for students' thinking*

Teacher attends to whether an answer or procedure is (or is not) correct.	Teacher identifies specific strategies or representations that are important to notice; strategically uses observations, student responses to questions, and written work to determine what students understand; and uses these data to inform in-the-moment discourse and future lessons.

Tool 4.1 Reflection Questions

1. What do you notice, in general, about your self-assessment of these *Shifts in Classroom Practice*?

2. What might be specific teaching moves that align with where you placed yourself on the *Shifts*?

3. What might be specific teaching moves that align *to the right of* where you placed yourself on the *Shifts*?

4. What might be some professional learning opportunities to help you move to the right for one or more of these *Shifts*?

4.2 Total Participation Technique (TPT) Planning
*(based on Himmele & Himmele, 2017)

Instructions to the Coach: Provide the Overview section of this chapter for teachers to read about TPT Cognitive Model prior to using this tool. This tool can be used in professional learning, PLCs, or a coaching cycle to consider how to devote more instructional time to Quadrant 4.

Instructions: Identify specific strategies and techniques focused on Quadrant 4 of the TPT Cognitive Engagement Model (see Coach's Digest for an explanation of the model). Not all activities have to be in Quadrant 4, but the majority should reside there to maximize learning.

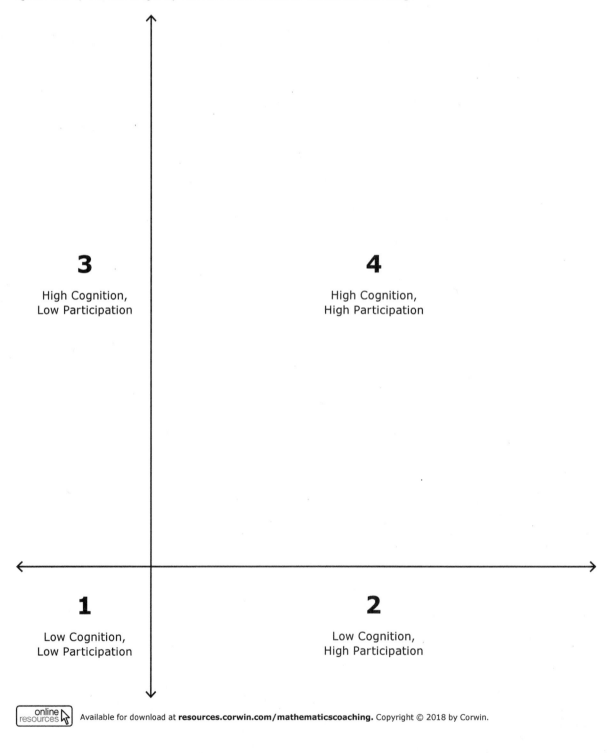

3

High Cognition,
Low Participation

4

High Cognition,
High Participation

1

Low Cognition,
Low Participation

2

Low Cognition,
High Participation

4.3 Planning for Cooperative Groups

Instructions to the Coach: In addition to using this as part of a planning cycle, each one of these categories can become a poster on which teachers brainstorm options for a selected task/lesson. This can also be used as a focus of a lesson study (see Tool 4.5 for data gathering).

Instructions: Cooperative groups are effective when they support positive social interactions (and not effective when they don't). Record ideas in the space provided to ensure your cooperative groups will foster positive social interactions.

Organization	*Accountability*
How will you organize groups? • How many students in a group? • Is there an ability mix? • Is there a gender mix? • Do students get to choose their group?	How will you build in individual accountability within group work? • Group roles? • Instructional strategies (e.g., Jigsaw)? • Other?
Your plan:	**Your plan:**
Management	*Assessment*
How will you organize the room to facilitate cooperative groups? • Transitions from whole group to cooperative groups? • Your role while students work in groups? • What materials do students need?	How will you assess individual students? Informal interviews while students work? • Checklist? • Anecdotal recording sheet? • Completed work? • Other?
Your plan:	**Your plan:**

Source: Tool previously published by Bay-Williams, J., McGatha, M., Kobett, B., and Wray, J. (2014). Mathematics Coaching: Resources and Tools for Coaches and Leaders, K–12. *New York, NY: Pearson Education, Inc.*

4.4 Engagement Techniques

Instructions to the Coach: Collect evidence of the teacher's use of engagement techniques. In the evidence column, record only data (what you see or hear), no judgments. During a reflecting conversation, collaboratively classify each activity according to the quadrant from the TPT Cognitive Engagement Model (see the Coach's Digest for an explanation of the model).

Evidence From the Lesson	Quadrant
Physical Activity	
Social Activity	

4.5 Cooperative Groups Data Gathering

Instructions to the Coach: Use this tool to collect data when a teacher is using cooperative groups. Record only data (what you see or hear), no judgments. If using this in a lesson study, be sure participants understand the difference between evidence and judgments.

Organization	Accountability

Management	Assessment

Source: Tool previously published by Bay-Williams, J., McGatha, M., Kobett, B., and Wray, J. (2014). Mathematics Coaching: Resources and Tools for Coaches and Leaders, K–12. New York, NY: Pearson Education, Inc.

4.6 Engagement Techniques Discussion Prompts

Instructions: Use any or all of these questions to reflect on students' engagement (and learning) in a lesson.

1. What opportunities did students have to reflect on their mathematical thinking individually and with others?

2. How did you incorporate opportunities for students to physically move during the lesson?

3. What did you do to facilitate that physical movement so it was productive?

4. How did your use of engagement techniques go compared to how you thought it would? Which strategies seemed most effective? Why?

5. What are some of the things you did that impacted the success of the engagement techniques you used in this lesson?

6. How do you think the students felt about the engagement techniques you used in the lesson?

7. How did the engagement techniques support the lesson objectives?

8. How did the engagement techniques support students demonstrating the Mathematical Practices?

9. How did the engagement techniques keep students actively engaged throughout the lesson?

10. How did the engagement techniques support you in determining if each student learned the objectives?

11. What are you learning about engagement techniques?

4.7 Reflecting on Cooperative Groups

Instructions to the Coach: This reflection tool can be used as a follow-up to the Cooperative Groups Planning and Data-Gathering Tools (4.3 and 4.5). Ask the teacher to bring Tool 4.3 and use it to reflect on each aspect of cooperative groups. Then, share the data you collected. Follow up with the reflection questions.

Instructions: In each box, record ideas related to what went well and what could be more effective.

Organization	Accountability
How will you organize groups? • How many students in a group? • Is there an ability mix? • Is there a gender mix? • Do students get to choose their group?	How will you build in individual accountability within group work? • Group roles? • Instructional strategies (e.g., Jigsaw)? • Other?
Your reflection:	**Your reflection:**
Management	Assessment
How will you organize the room to facilitate cooperative groups? • Transitions from whole group to cooperative groups? • Your role while students work in groups? • What materials do students need?	How will you assess individual students? • Informal interviews while students work? • Checklist? • Anecdotal recording sheet? • Completed work? • Other?
Your reflection:	**Your reflection:**

Source: Tool previously published by Bay-Williams, J., McGatha, M., Kobett, B., and Wray, J. (2014). Mathematics Coaching: Resources and Tools for Coaches and Leaders, K–12. *New York, NY: Pearson Education, Inc.*

4.8 Analyzing Learning in Cooperative Groups

Instructions to the Coach: This reflection tool can be used as a follow-up to the Cooperative Groups Planning and Data-Gathering Tools (4.3 and 4.5). It can be used with Tool 4.7 or on its own.

Instructions: Use any or all of these questions to reflect on the impact of cooperative groups on individual student learning.

Opening Questions
1. How did the cooperative group lesson go compared to how you thought it would?
2. What do you notice about the data I collected for you? How does it compare to your recollections of the lesson?
3. What criteria did you use to organize the groups? How did that work in terms of supporting student learning?

Learning of Individuals in the Group
4. What are some ways you built in individual accountability (each person is responsible to produce something)? How did that go? How did students respond?
5. What are some ways you built in shared responsibility (interdependence is built into the task)? How did that go? How did students respond?
6. In what ways did your management of the cooperative groups support students' individual learning?
7. How did your choice of assessment strategy provide you with useful data on individual students? How did it support students' learning?

Next Steps
8. What are you learning about cooperative groups?
9. What will you do again? What will you do differently?

Source: Tool previously published by Bay-Williams, J., McGatha, M., Kobett, B., and Wray, J. (2014). Mathematics Coaching: Resources and Tools for Coaches and Leaders, K–12. New York, NY: Pearson Education, Inc.

Chapter 5

Questioning and Discourse

Dear Coach,

Here is support for you as you work with teachers on questioning and discourse.

In the **COACH'S DIGEST** ...

Overview: A review of high-level questioning and facilitating discourse, including a discussion of talk moves. Also available as a download to share with teachers.

Coaching Considerations for Professional Learning: Ideas for how to support a teacher or a group of teachers in learning about questioning and discourse.

Coaching Lessons From the Field: Mathematics coaches share how they use a tool or idea related to support questioning and discourse.

Connecting to the Framework: Specific ways to connect selected *Shifts* and

Mathematical Practices to questioning and discourse.

Coaching Questions for Discussion: Menu of prompts for professional learning or one-on-one coaching about questioning and discourse.

Where to Learn More: Articles, books, and online resources for you and your teachers on where to learn more about questioning and discourse!

In the **COACH'S TOOLKIT** ...

Ten tools focused on questioning and discourse, for professional learning or coaching cycles.

Coach's Digest

In the Coach's Digest, we begin with an overview of questioning and **discourse**, written to teachers (and to you, the coach). As you read the Overview, the following questions might help you reflect on this topic in terms of your role as a mathematics coach:

- How might I use Bloom's Revised Taxonomy to help teachers ask good questions during instruction?
- How might I provide experiences for teachers in analyzing the patterns of their questions during a lesson?

Overview of Questioning and Discourse

Crafting and posing questions are both critical aspects of a teacher's practice, which is why "Pose purposeful questions" is one of the eight Effective Mathematics Teaching Practices in *Principles to Actions: Ensuring Mathematical Success for All* (NCTM, 2014). The first part of posing a question is to craft one that has the potential to elicit high-level thinking. The second aspect, posing the question, may seem like a routine act, but it also requires purposeful actions. In fact, crafting and posing questions both require explicit attention on the part of the teacher in order for this complex process to be effective.

Crafting High-Level Thinking Questions

Teachers ask many questions every day. Research from the early 1980s suggested that teachers asked between 300 and 400 questions daily (Leven & Long, 1981), a number that is probably even greater today. However, most questions asked by teachers do not support students in high-level thinking. Mid-continent Research for Education Learning (McREL) collected data from 23,000 classroom observations and found that 60 percent of questions posed were at the lowest two levels of Bloom's Taxonomy (McREL, 2009). Stigler and Hiebert (2009), focusing specifically on mathematics classrooms, found that middle school mathematics teachers in the United States posed fewer high-level questions than teachers in other countries.

Bloom's Taxonomy is probably the best-known framework for identifying the level of cognitive demand of questions (Bloom, Engelhart, Furst, Hill, & Krathwohl, 1956). In 2001, a revised version of Bloom's Taxonomy was published in an attempt to update the taxonomy to be relevant in the twenty-first century (Krathwohl, 2002; see Figure 5.1). Asking students questions from the upper end of these frameworks (Application/Applying, Analysis/Analyzing, Synthesis/Evaluating, and Evaluation/Creating) supports students in high-level thinking.

| Figure 5.1 | Bloom's Taxonomy |

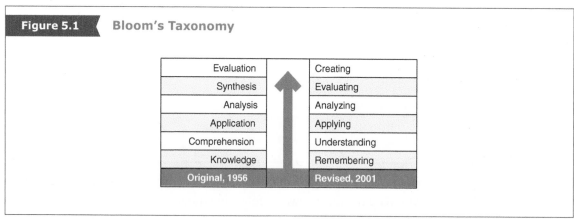

Original, 1956	Revised, 2001
Evaluation	Creating
Synthesis	Evaluating
Analysis	Analyzing
Application	Applying
Comprehension	Understanding
Knowledge	Remembering

Source: Krathwohl, D. R. (2002). A revision of Bloom's Taxonomy: An overview. *Theory Into Practice, 41*(4), 212–218.

Crafting high-level thinking questions for the lesson's learning objectives requires preplanning. To support teachers in planning effective questions, it can be helpful to consider the *purposes* questions might serve. These purposes include: (1) gathering information, (2) probing thinking, (3) making the mathematics visible, (4) encouraging reflection and justification, and (5) engaging with the reasoning of others (Huinker & Bill, 2017; NCTM, 2014, pp. 36–37). Once high-level thinking questions are crafted, the focus can change to posing questions in ways that ensure that *all* students are thinking.

Posing High-Level Thinking Questions

An important aspect of posing high-level questions is thinking about the *patterns* of questions. There are (at least) three patterns of questioning that typically occur in mathematics classrooms (Herbel-Eisenmann & Breyfogle, 2005). Most common is the **initiation–response–feedback** (IRF), in which the teacher asks a question ("What is … ?"), a student responds ("Fifteen"), and the teacher provides or evaluates the response ("Good"). Typically, this questioning pattern does not engage students in high-level thinking. A second questioning pattern is **funneling**, wherein the teacher leads students through a series of questions to the teacher's desired end. In this pattern, the teacher is doing the high-level thinking of making connections; the students are merely supplying quick responses to questions as the teacher pushes toward a conclusion. The third pattern, **focusing**, is a subtle but significant shift in questioning in which the teacher asks questions based on the students' thinking to support them in thinking at high levels. For example, after hearing a student respond, a teacher might say, "Tell me more about why you …" or "Look at Amy's equation and tell me how each variable and number connects to the story situation." Questions such as these probe into student thinking and ask them to make predictions, compare, classify, evaluate, analyze, or estimate.

Classroom Discourse

Posing effective questions is just one aspect of a larger concept: **classroom discourse** (also one of the NCTM Effective Mathematics Teaching Practices). The role of the teacher in supporting meaningful mathematical discourse is complex (Chapin, O'Connor, & Anderson, 2013; Hufferd-Ackles, Fuson, & Sherin, 2015; NCTM, 2014; Smith & Stein, 2011). Discourse involves various teaching actions, including the following:

- Asking questions to understand and deepen students' thinking
- Listening to students' responses to gauge their learning
- Encouraging students to listen and respond to their peers
- Requiring students to explain their thinking
- Encouraging students to use multiple representations
- Allowing students to engage in productive struggle

Chapin and colleagues (2013) identify **talk moves** and tools that teachers can use to orchestrate classroom discussions that support increased student participation, as seen in Figure 5.2(a). These talk moves are used in various discourse situations, illustrated by the example scenarios in Figure 5.2(b). Taking time to focus on crafting and posing high-level thinking questions is critical to facilitating productive discourse; making this a focus of collaborative work in PLCs can save everyone time. And the increased student involvement in classroom discourse will lead to increases in students' understanding of mathematics.

Figure 5.2 | Talk Moves and When to Use Them

5a. Talk Moves and Prompts

To understand students' contributions:

1. You used the hundreds chart, tell me more about that. (*Revoicing*)

2. How might you repeat [Ava's] thinking in your own words? (*Repeating*)

To deepen students' contributions:

1. How do you know that you are correct? (*Reasoning*)

2. What do other people think? (*Adding on*)

5b. Talk Move Map

You pose a question to the class.

Then, what if...

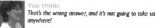

TALK MOVE MAP

⬇ **A STUDENT GIVES A RESPONSE.**

You think:
Huh?? I didn't understand that at all!

useful move:
▷ **Say more**

examples
Can you **say more about that**?
Could you **say that again**?
Can you **give an example** of what you mean?
So let me see if I understand. **Are you saying...**?

You think:
Gee, good point! Did everyone get that?

useful move:
▷ **Can someone rephrase or repeat that?**

examples
Can anybody **put that in their own words**?
Who thinks they **could repeat that**?

You think:
I think students got that, but I need to dig deeper into this student's thinking.

useful move:
▷ **Why do you think that?**

examples
What led you to **think about it that way**?
What's the **evidence** you used?
Can you **explain your reasoning** to us?
How did you **figure that out**?

You think:
Students heard this, but I want them to connect with this idea!

useful move:
▷ **What do other people think?**

examples
Who **agrees or disagrees**, and why?
Who wants to **add on** to what s/he just said?
What do you think about that idea?
Does anyone have a **different view**?

⬇ **FACES BLANK. ONLY 2 HANDS RAISED.**

You think:
I guess they need time to think!

▷ **Stop & Think** or **Stop & Jot** (60 seconds) *then*

▷ **Turn-and-talk** (60 seconds) *then*

▷ **Ask again!**

⬇ **A STUDENT GIVES A RESPONSE THAT IS WRONG OR CONFUSED.**

You think:
That's the wrong answer, but it might be very productive to discuss it!

Go back to the four moves to the left.

1. **Say more**
2. **Can someone rephrase that?**
3. **Why do you think that?**
4. **What do other people think?**

You think:
That's the wrong answer, and it's not going to take us anywhere!

▷ **Use your best judgment about how to move on.**

examples
Can you **say that again**?
Does anyone have a **different view**?
Well, actually, remember when we ... (give correction)

⬇ **SEVERAL STUDENT RESPONSES ARE OFF TOPIC.**

You think:
We've really gotten off track. Even though they're engaged, this isn't the question we're trying to consider!

▷ **Use your best judgment to get back on track.**

examples
Can you **link this back** to our question?
Can someone tell me how this fits in with our question?
Gee, what <u>was</u> our question? Let's recall where we're going...

Source: Developed by Cathy O'Connor and reprinted with permission of SERP.serpinstitute.org

Coaching Considerations for Professional Learning

Improving your skills at questioning and facilitating discourse is a career-long endeavor. It is important to help teachers distinguish among the many ideas surrounding questioning. For example, the following types of questions are often confused as the same:

- Open-ended
- High-level thinking
- Conceptually-focused

Yet questions can be open-ended and still low level. Or a high-level question can be focused on procedural knowledge. These aren't good–bad, right–wrong distinctions, but it is critical to make these distinctions when working on questioning so that everyone is focused on the same aspect of questioning. Here are some ideas for supporting teachers:

1. *Plan for questioning.* Planning for questioning can take time, but the effort is worth it in the long term. Making this the focus of a PLC can be a valuable way to collaborate to benefit all teachers (see Tools 5.2, 5.3, and 5.4). For example, focus on a unit and brainstorm misconceptions that students typically have related to the topic (see Smith & Stein, 2011). Then together, create high-level thinking questions that might address those misconceptions. As teachers use the questions in class, reflect on the outcomes in terms of students' thinking (see Tools 5.9 and 5.10).

2. *Analyze questions.* Volunteer to script questions during an observation (see Tool 5.5) and then work with the teacher to classify them in terms of Bloom's Taxonomy. Alternatively, provide time for a teacher to observe another or watch videos of classroom teaching to evaluate the types of questions and questioning patterns (see Coaching Lessons From the Field to see how one district did this).

3. *Track wait time.* A still prevalent issue related to asking questions is the amount of **wait time** teachers provide. Most questions are answered in less than one second (Rowe, 1972). Uncomfortable with silence, we jump in with an answer, a comment, a prompt, or another question. Increasing wait time to three to five seconds can lead to positive outcomes (Rowe, 1972; Stahl, 1994). Encourage your teachers to track their wait time by audio- or video-recording their lessons (see Tool 5.8). Alternatively, you can record wait times during an observation. Support teachers in creating strategies to increase their wait time and then track the changes in student thinking. Keene (2014) and Roake (2013) offer practical strategies for implementing wait time in the classroom (see Where to Learn More).

4. *Post questions.* Because good high-level thinking questions or conceptually-focused questions are tough to "wing," encourage teachers to post frequently-used good questions in the classroom. This has the added benefit of helping students pose questions of their peers (supporting Mathematical Practices such as 1, 2, and 3). Developing Mathematical Thinking With Effective Questions on PBS Teacherline (www.pbs.org/teacherline) provides a strong list. You can create bookmarks of these questions to give out to teachers and/or invite them to select key questions from this list to post in their rooms.

Coaching Lessons From the Field

In working with our teachers, we created a large "questioning grid" in their planning room based on Tool 5.10. During grade-level planning sessions, teachers visited classrooms to observe and record on sticky notes the first three questions they heard posed by the teacher. Each teacher analyzed the questions and placed them on the questioning grid and justified her or his thinking. The math coaches then selected some of the reflecting questions from Tool 5.10 to guide the group discussion.

The math coaches described the activity as an "eye-opener" in their schools because it allowed teachers to see that many of their questions were on the lower end of Bloom's Taxonomy. This activity supported teachers in focusing on asking high-level thinking questions.

—Sandra Davis, Rita Hays, Maria Leaman, Natasha Crissey, and Tracey Beck, Brevard Public School, Viera, FL

		Level of Thinking (Bloom's Taxonomy—Revised)					
		Remembering	Understanding	Applying	Analyzing	Evaluating	Creating
Mathematical Knowledge	Conceptual	• How would you explain the difference between probability and odds? • What is a negative number? • Which number is the denominator in this fraction? • What are the attributes of your shape?	• How many cubes do you think it will take to equal your rock? • Which of these triangles are similar? • What is the difference between supply and demand? • Which graph represents the data?	• Classify these triangles as scalene, equilateral, or obtuse.	• What are some ideas you have to classify these aliens?		• How could you create a playground given these parameters?
	Procedural	• What is the next step? • What is 12 take away 3? • What is the formula for the area of a circle? • What do you do first in this problem?	• Explain the steps for completing the order of operations. • How many lines do you need to draw?	• What is the perimeter of this triangle? • How much will it cost to have electricity, housing, and water?		• How might you know if your answer is correct?	

Connecting to the Leading for Mathematical Proficiency (LMP) Framework

As teachers focus on questioning and discourse, it is important for them to make explicit connections to the *Shifts in Classroom Practice* and the Mathematical Practices. The brief paragraphs that follow provide ideas for making these connections. You and your teachers can continue to add to these ideas! Tool 5.1 also focuses on connecting questioning and discourse to the *Shifts*.

Connecting Questioning and Discourse to *Shifts in Classroom Practice*

Shifts   1   2   3   **4**   **5**   6   **7**   **8**

Shift 4: From *show-and-tell* toward *share-and-compare*

Teacher has students share their answers.	Teacher creates a dynamic forum where students share, listen, honor, and critique each other's ideas to clarify and deepen mathematical understandings and language; teacher strategically invites participation in ways that facilitate mathematical connections.

Over recent years, many teachers have gotten better at asking students questions like "How did you solve it?" and "Who has solved it a different way?" This is a start, but questioning and discourse must not stop here if students are to think at a high level. Inserting questions such as "What do others think of this strategy?" "How do these strategies compare?" and "When might you use this strategy?" strengthens student understanding and helps students make mathematical connections. The talk moves (see Figure 5.2) and various tools can support this focus on share-and-compare (see Tools 5.3, 5.4, 5.5, 5.8 and 5.10).

Shift 5: From *questions that seek expected answers* toward *questions that illuminate and deepen student understanding*

Teacher poses closed and/or low-level questions, confirms correctness of responses, and provides little or no opportunity for students to explain their thinking.	Teacher poses questions that advance student thinking, deepen students' understanding, make the mathematics more visible, provide insights into student reasoning, and promote meaningful reflection.

A purposeful question strikes a delicate balance of being clear and being complex. A question might be clear but low level or focus on trivial mathematics. Or it can be complex but incomprehensible to students. A well-designed question, clear and high level, elicits important mathematical thinking and talking (see Tools 5.2, 5.3, 5.4, and 5.5). Focusing on the patterns of questions in a lesson can move teachers toward the right end of this *Shift* (see Tool 5.6).

Shift 7: From mathematics-made-easy toward mathematics-takes-time

Teacher presents mathematics in small chunks so that students reach solutions quickly.	Teacher questions, encourages, provides time, and explicitly states the value of grappling with mathematical tasks, making multiple attempts, and learning from mistakes.

Posing good questions is a great strategy when encouraging students to persevere in problem-solving. As students begin to struggle, a good question can provide the just-in-time support needed to encourage a student to keep going (see Tool 5.4). Asking students about their attempts, mistakes, and processes also communicates that doing mathematics is a journey, not just a destination—in other words, making multiple attempts, noticing and correcting errors, and continuing to seek a solution pathway is critically important in doing mathematics (it is not just about getting an answer).

Shift 8: From looking at correct answers toward looking for students' thinking

Teacher attends to whether an answer or procedure is (or is not) correct.	Teacher identifies specific strategies or representations that are important to notice; strategically uses observations, student responses to questions, and written work to determine what students understand; and uses these data to inform in-the-moment discourse and future lessons.

An important aspect of posing questions and productive classroom discourse is *listening* to students' responses. Using students' responses can inform in-the-moment classroom discourse as well as plan for future instruction. The distinctions between IRF, funneling, and focusing effectively capture this distinction (see Tool 5.6).

> Engage teachers in thinking about the *Shifts* related to questioning and discourse by having them think about specific teacher moves they can make along each of the continua (see Tool 5.1). For example, in *Shift 5,* the focus is on high-level questioning. After using Tool 5.5 to script questions for the teacher, follow up by placing the questions from the lesson along *Shift 5*. If questions tend to fall toward the middle or left end of the *Shift*, consider with the teacher what adjustments can be made so that the questions lie more to the right on the continuum.

NOTES

_____

_____

_____

_____

_____

_____

_____

_____

Connecting Questioning and Discourse to Mathematical Practices

MPs 1 2 3 4 5 6 7 8

One of the purposes of asking high-level thinking questions is to model for students the kind of thinking in which they need to engage. The goal is for students to be self-directed in asking themselves these questions as they solve mathematics tasks. Several of the Mathematical Practices include opportunities for students to engage in high-level thinking by questioning themselves and others (see Chapter 12 for a PD activity matching high-level questions with all of the Mathematical Practices).

1. *Make sense of problems and persevere in solving them.* In order for students to make sense of problems, they need to know the right questions to ask. Students learn how to ask questions of themselves when that has been modeled for them by their teachers. Asking questions such as the ones listed here can assist students in making sense of problems and persevering when dealing with tasks they find difficult:

   - How is this task similar to a previous task you have completed?
   - How might you solve a simpler task to help with this?
   - What helped you be successful in solving the problem?

2. *Reason abstractly and quantitatively.* This Practice embodies Bloom's high-level thinking. When students are evaluating an equation or expression to determine whether it is equivalent to an equation or expression written in a different form, they are reasoning abstractly. When students contextualize, they are creating examples for the abstract equations. For students to become proficient in this Practice, significant use of high-level questions must be incorporated into solving problems. Examples include the following:

   - What is the relationship between the data/situation and the equation?
   - How are these answers alike? Different?
   - What might be an example that would fit this expression?

3. *Construct viable arguments and critique the reasoning of others.* The only way for this proficiency to be developed is if teachers pose focusing questions and facilitate discourse in ways that push students to describe their own solutions and evaluate or add to those shared by their peers. This involves analyzing and evaluating—two of Bloom's high-level thinking categories. Students can ask questions of their peers when assessing their work, such as

   - What was your thinking about using [a graph] to solve this task?
   - How did you get [that equation]?
   - How is that answer the same as [Nick's]?

   And teachers facilitate such questioning by asking such things as

   - What do the rest of you think about Ann's strategy?
   - Are these strategies different or alike?
   - Which of these strategies would you pick if you had a problem like this, but more difficult numbers? Why?

4. *Model with mathematics.* Determining a mathematical model for a situation requires analyzing the situation and determining the symbolic representation for the situation. Students need experiences connecting symbols to situations as early as kindergarten and extensively in middle school and high school. Questions focused on making these

connections support their ability to model with mathematics. Examples include the following:

- How does your model (equation) connect to the situation?
- Where can you find [the rate] in the situation? The table? The equation?
- Are these two equations equivalent? Which (if either) is more efficient?"

7. ***Look for and make use of structure.*** Focusing discussions can help students see structure in equations. For example, asking young children whether 3 + 5 is the same as 5 + 3 and asking older students to compare mathematical models can help them see the distributive property in action. Questions that focus on structure include these examples:

- How is this problem similar to another problem you have solved before?
- What patterns did you notice across these problems?
- What are possible answers (range of values) for this problem? What information tells you this?

 Ask your teachers to connect the Mathematical Practices to questioning and discourse by asking these questions:

- What connections do you see between high-level thinking and the selected Mathematical Practices?

- What adjustments have you made in your questioning that support students in demonstrating the selected Mathematical Practices?

- How does the discourse during your lessons influence the opportunities for students to engage in these Mathematical Practices?

NOTES

_____

_____

_____

_____

_____

_____

_____

_____

_____

_____

_____

_____

_____

_____

_____

Questions Related to the Focus Zone: Questioning and Discourse

1. Knowing the benefits of wait time, what might be some strategies you could use to increase your wait time?

2. How might you take a low-level question and adapt it to be a high-level question?

3. What are some ways to incorporate high-level thinking questions into homework? Into reviewing homework?

4. What do you know about students' misconceptions in _____ that could influence the questions you could use in this lesson?

5. What might be some questions you can plan for this lesson to address those misconceptions?

6. How might you monitor your questioning patterns?

7. How might you incorporate the talk moves into this lesson? What might be the benefits?

8. What is your thinking about orchestrating the discourse in this lesson?

9. In what ways do you monitor who participates in classroom discussions?

10. What strategies might you use to be sure *all* students are thinking of the answers to the questions you ask?

Questions Related to the LMP Framework

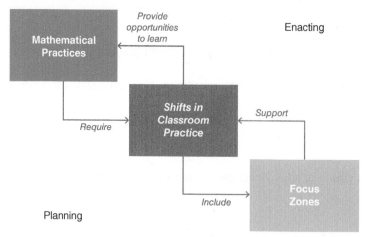

1. As you enact the ideas of questioning and discourse, what do you see as related *Shifts in Classroom Practice*?

2. As you enact the ideas of questioning and discourse, what do you see as opportunities to learn and, in particular, to develop the Mathematical Practices?

NOTES

_____

_____

_____

_____

_____

_____

_____

_____

_____

_____

_____

_____

_____

_____

_____

_____

_____

Where to Learn More

Books

Chapin, S. H., O'Connor, C., & Anderson, N. C. (2013). *Talk Moves: A Teacher's Guide for Using Classroom Discussions in Math* (3rd ed.). Sausalito, CA: Math Solutions.

The multimedia third edition of this popular book includes the talk moves and tools teachers can use in classroom discussions to support student learning in mathematics. New to this edition are 46 video clips showing math discussions in grades K–6 and more than 20 lesson plans ready for classroom use.

Francis, E. M. (2016). *Now That's a Good Question! How to Promote Cognitive Rigor Through Classroom Questioning*. Alexandria, VA: ASCD.

The author suggests eight different kinds of questions to support students in thinking at high levels and communicating depth of knowledge. Examples of good questions are provided across content areas and grade levels.

Parrish, S. (2014). *Number Talks: Whole Number Computation, Grades K–5: A Multimedia Professional Development Resource*. Sausalito, CA: Math Solutions.

This book explains the power of number talks and how they can support students' development of computation strategies. Guidance is provided on designing number talks, asking questions that build understanding, and developing grade-level-specific computation strategies. A facilitator's guide and streaming video clips of numbers talks in action are included.

Smith, M., & Stein, M. (2011). *Five Practices for Orchestrating Productive Mathematics Discussions*. Reston, VA: NCTM.

This book outlines a framework for orchestrating mathematically productive discussions based on student thinking. The five instructional practices are described and illustrated with examples from real classrooms. These practical suggestions will support teachers in orchestrating productive mathematics discussions that are responsive to students' thinking. A professional development guide is included.

Articles

Bahr, D. L., & Bahr, K. (2017). "Engaging All Students in Mathematical Discussions." *Teaching Children Mathematics, 23*(6), 350–359.

This article presents four strategies, aligned with NCTM's Effective Teaching Practice of "Pose purposeful questions," to ensure that students are actively engaged in mathematical discussions. The authors synthesized three different taxonomies to create a framework of cognitive complexity that includes the thinking level, a definition, and sample questions.

Ghousseini, H., Lord, S., & Cardon, A. (2017). "Supporting Math Talk in Small Groups." *Teaching Children Mathematics, 23*(7), 422–428.

This article highlights how to strategically use the launch phase of a lesson to support students' mathematical talk in small groups. The authors share three specific strategies with examples: (1) modeling for students what collaboration looks like in action, (2) creating opportunities for guided math talk, and (3) offering resources that support self-directedness in students.

Hodge, L. L., & Walther, A. (2017). "Building a Discourse Community: Initial Practices." *Mathematics Teaching in the Middle School, 22*(7), 430–437.

> *Hodge and Walther describe four initial practices that preservice teachers found beneficial in promoting mathematical discourse. They define each practice and provide examples of what the practice would look like "in action" using classroom tasks and discourse samples. These four practices provide foundational steps in creating an environment for a discourse community.*

Keene, E. O. (2014). "All the Time They Need." *Educational Leadership, 72*(3), 66–71.

> *The author describes the benefits of wait time and shares practical tips for teachers. Although written from the perspective of a language arts classroom, the tips apply to all content areas.*

Reinhart, S. (2000). "Never Say Anything a Kid Can Say!" *Mathematics Teaching in the Middle School, 5*(8), 478–483.

> *Reinhart describes his journey of going from a teacher that "explains things well" to a teacher that "gets kids to explains things well." He provides a useful list of questioning strategies that he developed as he transformed his teaching and created a classroom environment where students were actively engaged in learning mathematics.*

Roake, J. (2013). "Planning for Processing Time Yields Deeper Learning." *Education Update, 55*(8), 1, 6–7.

> *Roake presents 10 tips for building think time into classroom discussions. She provides a brief overview of the importance of processing time with connections to brain research and then outlines the 10 practical strategies. A quick but powerful read!*

Online Resources

Kathy Schrocks's Bloomin' Apps

www.schrockguide.net/bloomin-apps.html

> *This webpage has a collection of online apps aligned with the revised Bloom's Taxonomy.*

Levels of Cognitive Demand

http://mdk12.org/instruction/curriculum/mathematics/cognitive_levels.html

> *Organized around Bloom's Taxonomy, this tool provides competencies and skills aligned with the levels of cognitive demand, as well as various question cues to help promote high-level thinking skills.*

Developing Mathematical Thinking With Effective Questions on PBS Teacherline

http://www.pbs.org/teacherline

> *This resource provides questions for teachers to ask focused on various mathematical processes to support students when solving math tasks.*

Coach's Toolkit

These tools are a menu from which you can select any that make sense for your setting/context. They can be used independently or as part of a coaching cycle. You may start with the self-assessment, which can guide you in deciding which of the other tools may be most useful. If using these tools for a coaching cycle, mix and match as you like or use one of the combinations we suggest in the diagrams that follow. The tools in this chapter include instructions to the coach and the teacher. You can download copies of the tools that only have instructions for the teacher at **resources.corwin.com/mathematicscoaching.**

Self-Assess

5.1	Connecting *Shifts* to Questioning and Discourse Self-Assessment

Gather Data

5.5	Bloom's Taxonomy (Revised) and Mathematical Knowledge
5.6	Questioning Patterns
5.7	Wait Time
5.8	Productive Discussion and Talk Moves

Plan

5.2	High-Level Thinking Questions
5.3	Questioning Across Lesson Phases
5.4	Questioning Across Lesson Phases (Focus on Productive Struggle)

Reflect

5.9	Question and Discourse Discussion Prompts
5.10	Reflecting on Bloom's Taxonomy (Revised) and Mathematical Knowledge

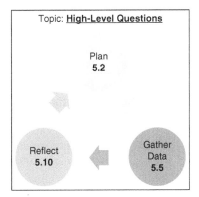

Additional Tools in Other Chapters

High-level thinking is also a focus in Chapter 2 (Implementing Effective Instruction; see Tool 2.5); Chapter 3 (Content Knowledge and Worthwhile Tasks; see Tools 3.2 and 3.5); Chapter 4 (Engaging Students; see Tool 4.2); Chapter 6 (Formative Assessment; see Tools 6.2 and 6.3); and Chapter 7 (Analyzing Student Work; see Tool 7.4).

 To download the coaching tools for Chapter 5 that only have instructions for the teacher, go to **resources.corwin.com/mathematicscoaching.**

5.1 Connecting *Shifts* to Questioning and Discourse Self-Assessment

Instructions to Coach: Ask teachers (individually or as part of a PLC activity) to self-assess where they position themselves on each of these Shifts in Classroom Practice *related to questioning and discourse. Use the questions that follow during a coaching conversation or in a PD setting to support teachers (and to help you decide which tools may be most useful from this chapter). A one-page version of this tool without this note is available for download.*

Instructions: The following *Shifts in Classroom Practice* have specific connections to questioning and discourse. Put an X on the continuum of each *Shift* to identify where you currently see your practice.

Tool 5.1 Shifts

Shift 4: From *show-and-tell* toward *share-and-compare*

Teacher has students share their answers.	Teacher creates a dynamic forum where students share, listen, honor, and critique each other's ideas to clarify and deepen mathematical understandings and language; teacher strategically invites participation in ways that facilitate mathematical connections.

Shift 5: From *questions that seek expected answers* toward *questions that illuminate and deepen student understanding*

Teacher poses closed and/or low-level questions, confirms correctness of responses, and provides little or no opportunity for students to explain their thinking.	Teacher poses questions that advance student thinking, deepen students' understanding, make the mathematics more visible, provide insights into student reasoning, and promote meaningful reflection.

Shift 7: From *mathematics-made-easy* toward *mathematics-takes-time*

Teacher presents mathematics in small chunks so that students reach solutions quickly.	Teacher questions, encourages, provides time, and explicitly states the value of grappling with mathematical tasks, making multiple attempts, and learning from mistakes.

Shift 8: From *looking at correct answers* toward *looking for students' thinking*

Teacher attends to whether an answer or procedure is (or is not) correct.	Teacher identifies specific strategies or representations that are important to notice; strategically uses observations, student responses to questions, and written work to determine what students understand; and uses these data to inform in-the-moment discourse and future lessons.

Tool 5.1 Reflection Questions

1. What do you notice, in general, about your self-assessment of these *Shifts in Classroom Practice*?
2. What might be specific teaching moves that align with where you placed yourself on the *Shifts*?
3. What might be specific teaching moves that align *to the right of* where you placed yourself on the *Shifts*?
4. What might be some professional learning opportunities to help you move to the right for one or more of these *Shifts*?

5.2 High-Level Thinking Questions

Instructions to the Coach: This tool is a three-step process and can be used in professional learning or a coaching cycle. In a coaching cycle, a teacher could complete it prior to or during a planning conversation.

Instructions: First, identify high-level thinking questions for each learning target in a lesson or unit. Second, consider anticipated challenges or misconceptions related to the learning goals and related questions. Third, add any additional questions that will address these challenges.

Learning Goal/Objective	High-Level Thinking Questions to Promote Discourse

Anticipated Student Challenges and/or Misconceptions

Additional High-Level Questions to Address Student Challenges and/or Misconceptions

5.3 Questioning Across Lesson Phases

Instructions to the Coach: Use this tool for a collaborative lesson study or for a planning conversation in a coaching cycle. You can download this and laminate it for repeated use or reference.

Instructions: Questions vary with phases of a lesson. Use this template to plan questions that might be appropriate to pose in an upcoming lesson.

Launching the Task
• What is the task asking you to do? • What do you already know about this topic? • What information do you have? What do you need to find out? • What strategies might you use to solve this problem? • What diagram, visual, manipulative, or table might you use to solve the problem? • What might your product (final solution) look like so that your classmates understand it?

Monitoring the Task [As students work]		
One-on-One	**Small Group**	**Whole Class**
• Where have you seen something like this before? • What might happen if I changed this part of the problem? • How is your strategy working? • What might be another way to think about this problem? • How might a simpler problem help you solve this problem? • How might a tool help you (number line, picture, manipulative)? • What patterns are you noticing? • Does your answer seem reasonable? Why or why not? *Question(s) focused on mathematics of the lesson (objective(s)):* _____ _____ _____ _____ _____ _____	Use one-on-one questions, plus … • What do you think of [group member's] strategy? • How are [two students in group] strategies alike or different? • Explain how [group member] solved the task. • How did you reach your conclusion(s)? • What might be a more efficient strategy? Or which of the strategies in your group are efficient? • Explain why you chose to organize your results this way. • Will this work with other numbers? Explain. • Are there other possibilities? How can you be sure? *Question(s) focused on mathematics of the lesson (objective(s)):* _____ _____ _____ _____	[To monitor thinking as students are still working] • What are some strategies you are using to solve the problem? • What have you noticed about this problem? • What do you think about what ____ said? • Do you agree? Why or why not? • Does anyone have the same answer but a different way to explain it? • Do you understand what ____ is saying? • Can you give me an example of ____? *Question(s) focused on mathematics of the lesson (objective(s)):* _____ _____ _____ _____ _____

Summarizing the Task—Whole Class
[To discuss task after students have solved it]

- How did you solve the problem?
- How might you convince the rest of us that your answer makes sense?
- Is that true for all cases or can you think of a counterexample?
- How does this relate to ___?
- What ideas that we have previously learned were useful in solving this problem?
- What would happen if ___? If ___ changes, how does it affect ___?
- What have you learned or found out today?
- What are the key points or big ideas in this lesson?

Question(s) focused on mathematics of the lesson (objective(s)):

_____

_____

_____

_____

_____

_____

_____

Question(s) focused on student solution strategies observed during the lesson

_____

_____

_____

_____

_____

_____

_____

_____

Source: Previously published by Bay-Williams, J., McGatha, M., Kobett, B., and Wray, J. (2014). Mathematics Coaching: Resources and Tools for Coaches and Leaders, K–12. New York, NY: Pearson Education, Inc.

5.4 Questioning Across Lesson Phases (Focus on Productive Struggle)

Instructions to the Coach: Use this tool for a collaborative lesson study or for a planning conversation in a coaching cycle. You can download this and laminate it for repeated use or reference, especially if working on Shift 7 and/or Mathematical Practice 1.

Instructions: Building perseverance requires support for students as they engage in productive struggle. Use this tool to plan strategies to support students so they will stay with a task. One tendency in teaching is to take away the struggle—this tool is targeted at *maintaining* the challenge but also at providing encouragement and structures to develop perseverance.

Anticipate student responses
How might students approach the problem (across the range of learners in your class)?

Launching the task
What explicit, specific statements might you make to **encourage** and **motivate** students to invest time in the task (not by telling them that it is relevant to a test or that they will need it for their future)?

Monitoring the task: Circle/highlight questions you might use. Add ones specific to the task(s).

To get started with the problem:	**If a student is stuck:**
• How would you describe the problem in your own words? • What information does the problem give you? • What strategies might you use? • What tools will you need? • What do you think the answer might be? • Others:	• Can you think of other problems like this one? • Could you try it with simpler numbers/shapes/situation? Fewer numbers? • Would it help to create a diagram? Make a table? Draw a picture? • Can you guess and check? • Others:
As students implement their strategies:	**If a student has finished (early):**
• Can you explain what you did so far? • Why did you decide to use this method? • Have you thought of all the possibilities? • Why did you decide to organize your results like that? • Do you see a pattern? • Others:	• What do you notice if … ? • Now that you are done, do you see another way (more efficient way) that you could have solved it? • Would your strategy work with other numbers/different parameters? • Others:

Summarizing the task
What explicit, specific statements might you make to **recognize** students' perseverance?

Source: Previously published by Bay-Williams, J., McGatha, M., Kobett, B., and Wray, J. (2014). Mathematics Coaching: Resources and Tools for Coaches and Leaders, K–12. *New York, NY: Pearson Education, Inc.*

 Available for download at **resources.corwin.com/mathematicscoaching**. Copyright © 2018 by Corwin.

5.5 Bloom's Taxonomy (Revised) and Mathematical Knowledge

Instructions to the Coach: Gather data for a teacher about high-level questions asked in each phase of a lesson. This tool can be used in connection with Tools 5.4 or 5.5. Script the questions and during a reflecting conversation, let the teacher identify the level of thinking from Bloom's Revised Taxonomy (Krathwohl, 2002) and the mathematical knowledge for each question.

Mathematical Knowledge Key	*Question Level of Thinking Key*	
CO: *Conceptual* P: *Procedural*	CR: *Creating* E: *Evaluating* AN: *Analyzing*	AP: *Applying* U: *Understanding* R: *Remembering*

Launching the Task

Monitoring the Task

Summarizing the Task

Source: Previously published by Bay-Williams, J., McGatha, M., Kobett, B., and Wray, J. (2014). Mathematics Coaching: Resources and Tools for Coaches and Leaders, K–12. *New York, NY: Pearson Education, Inc.*

5.6 Questioning Patterns

Instructions to the Coach: Collect data about the questioning patterns you notice the teacher using. Just script the questioning pattern (questions and responses) and let the teacher identify the pattern during a reflecting conversation. This tool could be used in conjunction with reading the article "Questioning Our Patterns of Questioning" (Herbel-Eisenmann & Breyfogle, 2005).

Teacher Questions	Student Responses	Questioning Patterns
		☐ IRF ☐ Funneling ☐ Focusing
		☐ IRF ☐ Funneling ☐ Focusing
		☐ IRF ☐ Funneling ☐ Focusing
		☐ IRF ☐ Funneling ☐ Focusing
		☐ IRF ☐ Funneling ☐ Focusing
		☐ IRF ☐ Funneling ☐ Focusing
		☐ IRF ☐ Funneling ☐ Focusing

Source: Previously published by Bay-Williams, J., McGatha, M., Kobett, B., and Wray, J. (2014). Mathematics Coaching: Resources and Tools for Coaches and Leaders, K–12. New York, NY: Pearson Education, Inc.

5.7 Wait Time

Instructions to the Coach: Script the questions a teacher asks and record the wait time before a response. Indicate whether the teacher or a student offers the response. This tool could be used in conjunction with either or both of these articles: "Planning for Processing Time Yields Deeper Learning" (Roake, 2013) and "All the Time They Need" (Keene, 2014).

Teacher Question	Wait Time (in seconds)	Response (T or S)

Source: Previously published by Bay-Williams, J., McGatha, M., Kobett, B., and Wray, J. (2014). Mathematics Coaching: Resources and Tools for Coaches and Leaders, K–12. *New York, NY: Pearson Education, Inc.*

5.8 Productive Discussions and Talk Moves

Instructions to the Coach: Use the goals and talk moves as a structure for collecting evidence during a lesson. Keep in mind that not all talk moves may be appropriate for every lesson.

Talk Moves	Teacher Questions/Statements	Student Responses
Goal 1: Help individual students share, expand, and clarify their own thinking.		
1. Time to think 2. Say more 3. So are you saying … ?		
Goal 2: Help students listen carefully to one another.		
4. Who can rephrase or repeat?		
Goal 3: Help students deepen their reasoning.		
5. Ask for evidence or reasoning 6. Challenge or counterexample		
Goal 4: Help students think with others.		
7. Agree/disagree and why? 8. Add on 9. Explain what someone else means		

Source: Adapted from goals and talk moves from Chapin, S. H., O'Connor, C., & Anderson, N. C. (2013). Talk Moves: A Teacher's Guide for Using Classroom Discussions in Math (3rd ed.). Sausalito, CA: Math Solutions.

5.9 Question and Discourse Discussion Prompts

Instructions to the Coach: Use these reflecting questions to discuss any or all of the questioning strategies identified within brackets in the questions below. For example, the entire discussion can focus on one element (e.g., talk moves) or can focus on several elements.

Instructions: Reflect on your use of questioning and discourse using these prompts.

1. What was your awareness of [questioning patterns, high-level thinking questions, talk moves, wait time]? (Hand the teacher the data you collected after he or she responds to this question and provide time for him or her to study the data.)

2. What do you notice about the data?

3. How did your [questioning patterns, high-level thinking questions, talk moves, wait time] go compared to how you thought it would?

4. How did your [questioning patterns, high-level thinking questions, talk moves, wait time] impact students' thinking?

5. How did students react to your [questioning patterns, high-level thinking questions, talk moves, wait time]? What might account for this?

6. How did you make decisions about [questioning patterns, high-level thinking questions, talk moves, wait time]?

7. What are you learning about [questioning patterns, high-level thinking questions, talk moves, wait time]?

Source: Previously published by Bay-Williams, J., McGatha, M., Kobett, B., and Wray, J. (2014). Mathematics Coaching: Resources and Tools for Coaches and Leaders, K–12. *New York, NY: Pearson Education, Inc.*

5.10 Reflecting on Bloom's Taxonomy (Revised) and Mathematical Knowledge

Instructions to the Coach: This tool can be used for a one-on-one reflecting conversation, lesson study debrief, or PLC activity (using data from a video or the teacher's own lessons). For a coaching cycle, share the data you collected during the lesson (Tool 5.6). Allow the teacher time to identify the level of thinking from Bloom's Revised Taxonomy (Krathwohl, 2002) and the mathematical knowledge for each question. Using the question grid that follows, invite the teacher to indicate with tallies how many questions were in each cell. Use the questions as a guide to a reflecting conversation about the data.

Instructions: Use this grid to categorize questions from a lesson; then, discuss the follow-up questions.

		Level of Thinking (Bloom's Taxonomy—Revised)					
		Remembering	Understanding	Applying	Analyzing	Evaluating	Creating
Mathematical Knowledge	Conceptual						
	Procedural						

Questions about the coding of questions from the data-gathering tool:

1. What do you notice about the *mathematical knowledge of questions* posed in each phase of an inquiry lesson?

2. What do you notice about the *level of thinking of questions* posed in each phase of an inquiry lesson?

3. Which questions were most effective? Why?

Questions about the question grid:

4. What patterns do you notice in the question grid?

5. What questions might have strengthened the lesson? In other words, are there cells in the question grid that could have been asked (e.g., a conceptual question that involved application)?

6. What new questions might be developed in any of the cells in the question grid in preparing for the next lesson?

7. What might be some connections between particular levels of thinking questions and conceptual, procedural, or factual knowledge?

Source: Previously published by Bay-Williams, J., McGatha, M., Kobett, B., and Wray, J. (2014). Mathematics Coaching: Resources and Tools for Coaches and Leaders, K–12. New York, NY: Pearson Education, Inc.

Chapter 6

Formative Assessment

Dear Coach,

Here is support for you as you work with teachers on formative assessment.

In the **COACH'S DIGEST** ...

Overview: To review formative assessment and/or download and share with teachers.

Coaching Considerations for Professional Learning: Ideas for how to support a teacher or a group of teachers in learning about formative assessment.

Coaching Lessons From the Field: Mathematics coaches share how they used a formative assessment tool or idea from this chapter.

Connecting to the Framework: Selected *Shifts in Classroom Practice* and Mathematical Practices are connected to formative assessment.

Coaching Questions for Discussion: Menu of prompts for professional learning or one-on-one coaching about formative assessment.

Where to Learn More: Articles, books, and online resources for you and your teachers!

In the **COACH'S TOOLKIT** ...

Eleven tools focused on formative assessment, for professional learning or coaching cycles.

Coach's Digest

In the Coach's Digest, we begin with an overview of formative assessment, written to teachers (and to you, the coach). As you read the Overview, the following questions might help you reflect on this topic in terms of your role as a mathematics coach:

- What kinds of formative assessment practices do you see your teachers implementing with ease? Which formative assessment practices challenge your teachers?
- What types of professional learning activities, readings, and/or coaching cycles/lesson study might support their efforts to implement these ideas?
- What data might you collect from teachers and/or classrooms to evaluate teachers' formative assessment practices?

Overview of Formative Assessment

 To download the Chapter 6 Overview, go to **resources.corwin.com/mathematicscoaching.**

Formative assessment, also called *assessment for learning* (Stiggins, 2005) and *formative evaluation* (Hattie, 2009), focuses on using evidence to adjust instruction to positively impact student learning (Black & Wiliam, 1998). The National Council of Teachers of Mathematics (2014) states that "an excellent mathematics program ensures that assessment is an integral part of instruction, provides evidence of proficiency with important mathematics content and practices, includes a variety of strategies and data sources, and informs feedback to students, instructional decisions, and program improvement" (p. 5). Teachers need knowledge of both the overall purpose of formative assessment and its techniques that focus on planning, collecting, and using critical student evidence to provide **feedback**. Wiliam (2007) offers five key strategies that should guide the development and use of formative assessment in the classroom:

1. ***Clarifying, sharing, and understanding goals for learning and criteria for success with learners.*** Connected to the NCTM (2014) Teaching Practice that states "Establish mathematics goals to focus learning," teachers must continually clarify learning goals and the success criteria for students. Teachers should clearly identify and communicate the learning goals, which are statements about what the students are expected to learn and the learning activity. The learning goals have several purposes and may encompass knowledge, skills, practices, and concepts. Teachers need to be clear about why they are using particular learning activities, strategically connect those activities to the learning goals, and communicate to students, parents, and leaders how they will support students in mastering the learning goal (Hattie et al., 2016).

2. ***Engineering effective classroom discussions, questions, activities, and tasks that elicit evidence of students' learning.*** Eliciting evidence of students' learning requires careful attention to the use of formative assessment techniques that capture student understanding to inform instructional decision-making. Fennell, Kobett, and Wray (2017) describe specific formative assessment techniques that teachers can use to plan and facilitate formative assessment probes, collect evidence, and provide feedback. Brief descriptions of the **Formative 5** techniques follow:

   - *Observation.* Teachers observe students every day. The key factor is how they plan for, collect evidence from, and respond to those **observations**. As teachers anticipate what they will observe, they can also develop strategic feedback that supports student thinking about the mathematical concepts they are learning.
   - *Interview.* Teachers plan for and conduct brief (2–3 minutes) formative **interviews** to learn more about student thinking during the lesson. The individual, paired, or small-group interviews are orchestrated during the lesson as previously planned or in response to particular student strategies or mathematical understandings.
   - *Show Me.* The **Show Me** technique is a "performance response by a student or group of students that extends and often deepens what was observed and what might have been asked during the interview" (Fennell et al., 2017, p. 63). Teachers use Show Me prompts to gather important in-the-moment information about student understanding, particularly as they notice misconceptions or unique responses during the course of a lesson.

- *Hinge question.* Derived from Wiliam and Leahy (2015), the **hinge question** is a strategic question asked at a particular point during the lesson that provides a check for understanding and indicates to the teacher whether students are ready to move on or need more time to develop understanding.
- *Exit tasks.* The **exit task** is a fully developed rich task or problem that assesses the learning intention for the day, a cluster of learning intentions, or a standard. Far beyond what teachers know as exit slips or exit tickets, the exit task provides an opportunity for students to demonstrate what they know at a deeper level.

3. *Providing feedback that moves learning forward.* Teacher feedback is a critical formative assessment factor because teachers immediately address and respond to students during the lesson and advance their learning during the lesson. Feedback that is targeted to student learning goals, responsive to student needs, and provided in a timely manner is one of the teacher practices that has the strongest influence on students' learning (Hattie et al., 2016). As teachers plan to implement particular formative assessment techniques, they anticipate student responses to questions, probes, and tasks. Formative assessment techniques assist teachers in providing explicit and swift feedback to students because they have thought deeply about the range of student responses and mathematical understanding.

4. *Activating students as owners of their own learning.* Formative feedback naturally activates students to own their learning because they are able to respond and apply the feedback they receive from teachers. Teachers must purposely create opportunities for students to reflect upon the feedback they are receiving. This type of feedback can be written or verbal, but it must point students to self-evaluate or move toward the learning goals. Shepard (2008) points out the following characteristics of effective feedback:

- It directs attention to the intended learning, pointing out strengths and offering specific information to guide improvement.
- It occurs during learning, while there is still time to act on it.
- It addresses partial understanding.
- It does not do the thinking for the students.
- It limits corrective information to the amount of advice the students can act on.

Students who are engaged in monitoring and regulating their own learning process experience dramatic increases in learning (Fontana & Fernandes, 1994; Mevarech & Kramarski, 1997, as cited in Wiliam, 2007). Teachers can activate self-assessment by developing **rubrics** that help students self-assess their progress.

5. *Activating students as learning resources for one another.* Teachers who create a classroom community that is rich in collaboration share the responsibility of learning with all students and stimulate students to support each other during the learning process. Through robust discourse opportunities, students explain their own thinking and at the same time challenge each other to justify their reasoning. Within this process, students clarify their own understanding, push others' thinking, and remediate peers' misconceptions.

Clearly, formative assessment is complex! Supporting teachers in making this transformation in their practice must address these complexities and therefore takes time. Using Wiliam's (2007) five key strategies can help you effectively plan how to guide teachers on this important journey.

Coaching Considerations for Professional Learning

At the same time that teachers understand the compelling benefits of regular use of formative assessment strategies, they can become overwhelmed with the thought of integrating thoughtful formative assessment practices into their daily routine of planning and implementing lessons. Additionally, some teachers see formative assessment as simply engaging in a few actions in their toolkit, such as using exit slips or every-pupil response strategies (e.g., thumbs up/thumbs down), but they do not glean the full benefit of using formative assessment data to guide effective and purposeful teaching decisions. Here are some ideas for helping teachers maneuver the complexities of formative assessment:

1. ***Develop a formative assessment "big picture."*** Help teachers to understand the big ideas of formative assessment. Three critical resources that capture the importance of formative assessment and effectively describe the dramatic influence formative assessment has on student learning include the following:

   - The NCTM research brief *Five "Key Strategies" for Effective Formative Assessment* (www.nctm.org) explains the process of formative assessment (summarized in the Overview). This is a great place to start with helping teachers see the big picture of formative assessment (also see Tool 6.2).
   - *Formative 5: Everyday Assessment Techniques for Every Math Classroom* (Fennell et al., 2017) provides five formative techniques to guide teachers' planning and implementation of formative assessment. Examples, suggestions, and tools for all five techniques are included (see Tools 6.4, 6.6, and 6.7).
   - *Embedded Formative Assessment* (2011), written by Dylan Wiliam, translates research into practice and unpacks formative assessment as a bridge between teaching and learning.

2. ***Create a comprehensive formative assessment plan.*** Planning for formative assessment is critical, and it doesn't just happen! You can work with teachers in small groups or with an entire school to develop a **formative assessment plan** that incorporates these methods:

   - Techniques for planning formative assessment prompts (see Tools 6.3, 6.4, and 6.5)
   - Formative assessment data collection tools and student evidence (see Tools 6.6, 6.7, and 6.8)
   - Feedback targeted for particular mathematical understandings
   - Steps for teaching students about their role in hearing, using, and giving feedback
   - Daily short- and long-term goals

3. ***Consider online tools.*** There are a multitude of online tools that can help teachers capture student understanding. If teachers are using devices, consider using tools such as Classkick (www.classkick.com), Formative (www.goformative.com), and Explain Everything (www.explaineverything.com), to name just a few. These online tools can also be very useful in helping students and families showcase learning, understand feedback, and connect it to the learning intentions and success criteria.

4. ***Conduct diagnostic interviews.*** **Diagnostic interviews** (see Tools 6.8, 9.4, and 9.8), also called **clinical interviews**, are valuable in providing teachers with in-depth understanding of students' mathematical thinking because they move beyond what you might learn from student work (Ginsburg, 1997, 2009). Different than a brief interview described earlier, the diagnostic interview provides an opportunity for a teacher to learn about and diagnose student misconceptions as well as their advanced thinking. While teachers do not have

time to interview all students, even interviewing just a few of them can provide valuable insight to inform planning and instruction. Novice teachers, in particular, can enhance their understanding about how students think about the content they are teaching and plan appropriate interventions (Hodges, Rose, & Hicks, 2012).

Recording video or audio interviews and using them in professional learning can be a powerful learning experience (Jacobs, Ambrose, Clement, & Brown, 2006; Wright & Ellemor-Collins, 2008). Teachers can collectively decide on tasks for the interview and then interview one or two students each. When sharing the videos of the student interviews in your PLC, use a simple sharing **protocol** (Share, Observe, Suggest) to help teachers focus on the students' strategies and conceptual understanding. Follow up the sharing by creating questions that might further teachers' understanding of the students' thinking. For example, teachers may want to focus on correctness of answers rather than underlying thinking. Pose the new questions for the next round of interviews to probe teachers to think deeply about how students demonstrate their understanding. Finally, invite teachers to explore how the evidence from the interviews might inform instruction.

5. *Use rubrics.* Rubrics can be used as formative assessment tools to provide useful and specific feedback to students. Rubrics are most often used in a summative way, so teachers may need support in conceptualizing how rubrics can be used formatively (see Tools 6.4 and 6.10). Encourage teachers to involve students in the creation of rubrics (McGatha & Darcy, 2010), which supports students in thinking about the criteria for quality work (see Tool 6.4). It also provides them with a sense of ownership in the assessment process. Asking students to complete a rubric for their work on a particular task—then comparing that with a rubric completed by the teacher or another student—is a great way for students to become more accurate in their self-assessment and connect their learning to success criteria.

Coaching Lessons From the Field

Although this is the beginning of my second year as a coach, I am already leading our focus on formative assessment. Our school leadership agreed that we should focus our teachers on planning for formative assessment using a few key techniques (we are using the Formative 5), collecting evidence that will be shared during professional learning and reflecting on the kinds of feedback that teachers give to students as it connects to the evidence. I used *Shifts 5* and *8* as starting points for this work. After giving my teachers some information about the five key strategies, I asked them to reflect on where they were by placing a dot on the *Shifts* and then share their thoughts about where they wanted to go from there (see Tool 6.1).

Shift 8: From *looking at correct answers* toward *looking for students' thinking*

Teacher attends to whether an answer or procedure is (or is not) correct.	→	Teacher identifies specific strategies or representations that are important to notice; strategically uses observations, student responses to questions, and written work to determine what students understand; and uses these data to inform in-the-moment discourse and future lessons.

I am better at selecting strategies and representations to be shared, but still not sure what to do with the incorrect strategies that I see. I need to work on how to respond to errors in one-on-one and whole-class situations.

Connecting to the Leading for Mathematical Proficiency (LMP) Framework

As teachers focus on formative assessment, it is important for them to make explicit connections to the *Shifts in Classroom Practice* and the Mathematical Practices. The brief paragraphs that follow provide ideas for making these connections. You and your teachers can continue to add to these ideas (see Tool 6.1 for connecting to the *Shifts*)!

Connecting Formative Assessment to *Shifts in Classroom Practice*

Shifts: **1** 2 3 4 **5** 6 7 **8**

Shift 1: From *stating-a-standard* toward *communicating expectations for learning*

Teacher shares broad performance goals and/or those provided in standards or curriculum documents. →	Teacher creates lesson-specific learning goals and communicates these goals at critical times within the lesson to ensure students understand the lesson's purpose and what is expected of them.

Teachers use formative assessment practices to plan for moments within the lesson to collect evidence of student learning and provide explicit feedback to students that moves them toward learning goals. This feedback is provided to them during the learning process, when it is most powerful, so that both students and teachers can respond. Students are also given opportunities to self-reflect and self-assess on their learning alongside learning intentions and success criteria.

Shift 5: From *questions that seek expected answers* toward *questions that illuminate and deepen student understanding*

Teacher poses closed and/or low-level questions, confirms correctness of responses, and provides little or no opportunity for students to explain their thinking. →	Teacher poses questions that advance student thinking, deepen students' understanding, make the mathematics more visible, provide insights into student reasoning, and promote meaningful reflection.

As teachers design formative assessment prompts, they need to be mindful of questions that focus too quickly on producing answers, which will not reveal true mathematical understanding and may elicit surface understanding from students. *Careful planning is the very best antidote to low-level questioning.* Encourage teachers to first anticipate student responses to lesson activities and questions and then strategically plan for how they will respond.

Shift 8: From *looking at correct answers* toward *looking for students' thinking*

Teacher attends to whether an answer or procedure is (or is not) correct. →	Teacher identifies specific strategies or representations that are important to notice; strategically uses observations, student responses to questions, and written work to determine what students understand; and uses these data to inform in-the-moment discourse and future lessons.

With the very best intentions, teachers break down concepts into bite-sized chunks so that students can quickly grasp mathematical concepts. However, these bite-sized pieces often leave students struggling to make sense of mathematical big ideas. For example, teachers might break down the subtraction algorithm by asking students a series of questions (e.g., What do we do first? Second?). When this happens, formative assessment is used superficially to confirm what teachers hope to see. In this example, students dutifully subtract one column before another. Teachers note that students seem to know each part but may lose sight of the bigger concept of subtraction. To shift teachers to move toward looking for student thinking, have them unpack what student evidence would look like for a range of mathematical understandings within the concept. For our subtraction example, teachers would need to consider both conceptual understanding (what happens when students subtract) and procedural (what are the steps for subtracting) and then develop responses to student thinking.

> Ask your teachers to connect the *Shifts* to formative assessment by having them think about specific formative assessment practices that can move them along the continuum (see Tool 6.1). For example, you might ask each teacher to select a *Shift* and a key formative assessment strategy they could implement. A teacher might select *Shift 8* as their target and decide to design specific formative assessment tasks and prompts that will elicit students' use of representations.

NOTES

_____
_____
_____
_____
_____
_____
_____
_____
_____
_____
_____
_____
_____
_____
_____
_____
_____
_____
_____
_____
_____
_____
_____
_____
_____

Connecting Formative Assessment to Mathematical Practices

MP	**1**	**2**	**3**	4	5	6	7	**8**

At first glance, it may seem as if formative assessment would not be addressed in the Standards for Mathematical Practice. However, when you look a little closer, you will see that formative assessment, as described in the five key strategies (Wiliam, 2007) discussed previously, is embedded in several of the Standards for Mathematical Practice. We highlight four here.

1. ***Make sense of problems and persevere in solving them.*** An important part of formative assessment is student self-assessment (Key Strategy 4 in the Overview), which is addressed in Mathematical Practice 1. It is imperative that students understand how to monitor their work and make adjustments as needed. Helping students learn to monitor their work may take explicit **scaffolding** for some of them. For example, teachers can require students to examine misconceptions and incomplete understandings as (a) a simple mistake or (b) something they don't understand. If it is a simple mistake, students make the correction and identify what the error was. If they don't understand it, they will need to get feedback from the teacher or other students to assist them in identifying the mistake and making corrections. Teachers can also ask students to monitor work over the course of a unit by listing the learning intentions and success criteria for the entire unit and having them periodically self-evaluate whether they have achieved the learning intentions. Students can then develop a plan to determine the next steps for learning. There are many others ways to support students in learning to monitor their work that will help them to internalize the monitoring process.

2. ***Construct viable arguments and critique the reasoning of others.*** When considering formative assessment in terms of this Mathematical Practice, peer assessment comes to mind (see Key Strategy 5 in the Overview section of this chapter). Learning to critique the reasoning of others is a critical skill. Involving students in peer assessment benefits not only the students receiving the feedback but also the students providing the feedback. Using rubrics is one way to support students in peer assessment. In order to provide feedback using a rubric, the student must pay attention to the success criteria presented in the rubric and make judgments about whether the criteria have been met in the work they are reviewing. This is sometimes easier and less emotional for students to do with others' work than their own. We also know that sometimes students can communicate ideas more effectively to other students than teachers can.

3. ***Model with mathematics.*** Formative assessment can be used to determine whether students can make connections and move between contexts (mathematical and real world), visuals, and symbols, which is highlighted by Key Strategy 2 in the Overview section of this chapter. The focus of mathematical modeling is on the students' interpretation of the context (English, Fox, & Watters, 2005). Teachers can design tasks and use formative assessment techniques that probe students' thinking through observation and questioning to assess how students represent real-world scenarios with mathematical representations. For example, primary students might be asked to use the Show Me technique to record a number sentence to match a story, while older students would be asked to interpret real-world situations using mathematical ideas, formulas, and theorems in a robust exit task.

8. ***Look for and express regularity with reasoning.*** Mathematically proficient students are self-assessors (see Key Strategy 4 in the Overview), and explicit effort to incorporate self-assessment and reflection can support students in the development of this Mathematical Practice. Students are continually monitoring their own work and seeking patterns. They can also connect their current work with prior mathematics lessons and determine if their learning matches the success criteria. Teachers can prompt students to demonstrate that they are making sense of the mathematics and seeing connections by explaining their thinking in interviews and Show Me prompts.

→ Ask your teachers to connect the Mathematical Practices to formative assessment by asking the following questions:

- In what ways can you use formative assessment to gather information about how students exhibit these practices?

- What evidence can you collect about how students make sense of problems and persevere when solving, model with mathematics, and demonstrate their reasoning?

- How can you use this evidence to support students in engaging in the Mathematical Practices?

NOTES

_____

_____

_____

_____

_____

_____

_____

_____

_____

_____

_____

_____

_____

_____

_____

_____

_____

_____

_____

_____

Questions Related to the Focus Zone: Formative Assessment

1. Discuss ways to connect what teachers are currently doing to the five key strategies for effective formative assessment (Wiliam, 2007). Here are some example prompts:

   - What connections do you see between the activity and the learning goal for this lesson?
   - In what ways might you use formative assessment in this lesson/unit?
   - How will you anticipate students' responses to the questions, activities, and tasks that you designed?
   - How might you collect data about student learning during the lesson?
   - In what ways will you provide feedback to students?
   - How will you anticipate and identify misconceptions to move learning forward?
   - What might be the benefits of allowing students to self-assess their progress?
   - How might you use a rubric to activate students as owners of their learning?
   - How can you facilitate peer-to-peer feedback?
   - How might you support students in using peer assessment during learning?

2. As you consider the five key strategies, what do you consider your areas of strength and opportunities for growth? How could we move these into action steps?

3. What might be some ways to introduce and engage students with the five key strategies for formative assessment?

4. What Formative 5 techniques (or other techniques) might you use to incorporate the five key strategies into your practice?

5. How do you plan for, anticipate, and record student responses to learning activities?

Questions Related to the LMP Framework

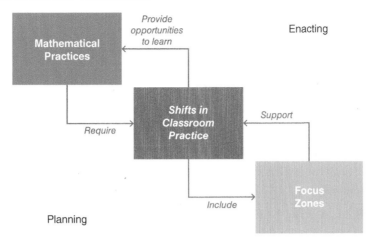

1. As you enact the ideas of formative assessment, what do you see as related *Shifts in Classroom Practice*?

2. As you enact the ideas of formative assessment, what do you see as opportunities to learn and, in particular, to develop the Mathematical Practices?

NOTES

_____

_____

_____

_____

_____

_____

_____

_____

_____

_____

_____

_____

_____

_____

_____

_____

_____

_____

_____

_____

_____

Where to Learn More

Books

Collins, A. M. (Ed). (2011). *Using Classroom Assessment to Improve Student Learning: Math Problems Aligned With NCTM and Common Core State Standards.* Reston, VA: NCTM.

> *Collins provides many examples of tasks that can be used as formative assessments in the middle grades. A professional development guide is included, as well as a discussion of classroom practices that align with formative assessment.*

Fennell, F., Kobett, B. M., & Wray, J. (2017). *Formative 5: Everyday Assessment Techniques for Every Math Classroom.* Thousand Oaks, CA: Corwin.

> *This book provides practical classroom-based techniques—observations, interviews, Show Me, hinge questions, and exit tasks that can be designed and implemented to inform instructional decision-making.*

Laud, L. (2011). *Using Formative Assessment to Differentiate Mathematics Instruction: Seven Practices to Maximize Learning.* Reston, VA: NCTM.

> *In addition to the many examples of formative assessment, including diagnostic interview questions, portfolio ideas, writing prompts, rubrics, checklists, and questions this book offers valuable ideas on how to* use *these assessments.*

Tobey, C. R., Minton, L. G., & Arline, C. B. (2006). *Uncovering Student Thinking in Mathematics: 25 Formative Assessment Probes.* Thousand Oaks, CA: Corwin.

Rose, C. M., Minton, L. G., & Arline, C. B. (2006). *Uncovering Student Thinking in Mathematics: 30 Formative Assessment Probes for the Secondary Classroom.* Thousand Oaks, CA: Corwin.

> *These two books provide a variety of assessment probes and tasks, including those that use manipulatives and representations.*

Wiliam, D. (2011). *Embedded Formative Assessment.* Bloomington, IN: Solution Tree.

> *Wiliam presents over 50 techniques and research evidence for the five key strategies presented in this chapter.*

Articles

Black, P., & Wiliam, D. (1998). "Inside the Black Box: Raising Standards Through Classroom Assessment." *Phi Delta Kappan, 80*(2), 139–148.

> *The authors of this seminal article provide a brief review of the formative assessment research evidence.*

Duckor, B., Holmberg, C., & Becker, J. R. (2017). "Making Moves: Formative Assessment in Mathematics." *Mathematics Teaching in the Middle School, 22*(6).

> *This article explores seven formative assessment moves that teachers can enact in combinations to uncover students' mathematical understanding.*

Fagan, E., Tobey, C. R., & Brodesky, A. (2016). "Targeting Instruction With Formative Assessment Probes." *Teaching Children Mathematics, 23*(3), 146–157.

> *The power of using formative assessment probes to collect data about student thinking and anticipate potential student misconceptions is effectively illustrated and described in this excellent article.*

Online Resources

Classkick

https://www.classkick.com

> *A free application to monitor students' responses to prompts and provide them with real-time feedback.*

Explain Everything

https://explaineverything.com

> *An application that can record student voice and representations, which can be stored and shared in a variety of ways.*

Formative

https://goformative.com

> *A free application to monitor students' responses to prompts in real time.*

NOTES

Exploring Zones on the Journey

Coach's Toolkit

These tools are a menu from which you can select any that make sense for your setting/context. They can be used independently or as part of a coaching cycle. You may start with the self-assessment, which can guide you in deciding which of the other tools may be most useful. If using these tools for a coaching cycle, mix and match as you like or use one of the combinations we suggest in the diagrams that follow. The tools in this chapter include instructions to the coach and the teacher. You can download copies of the tools that only have instructions for the teacher at **resources.corwin.com/mathematicscoaching.**

Self-Assess

6.1	Connecting *Shifts* to Formative Assessment Self-Assessment

Plan

6.2	Planning for Five "Key Strategies" for Formative Assessment
6.3	Developing Questions to Target Misconceptions
6.4	Using Rubrics for Formative Assessment
6.5	Observing and Providing Feedback

Gather Data

6.6	Observing Students' Thinking
6.7	Observing Students' Representations
6.8	Brief Formative Assessment Interview
6.9	Five "Key Strategies" for Formative Assessment Data Collection

Reflect

6.10	Using Rubrics for Formative Assessment
6.11	Analyzing Formative Assessment Key Strategies

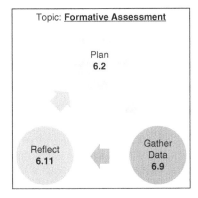

Additional Tools in Other Chapters

Tools in Chapter 7 (Analyzing Student Thinking) can also be used to support formative assessment work; Tools 7.4, 7.5, and 7.6 specifically examine student knowledge, strengths, and misconceptions. Tools in Chapters 9 include diagnostic interview tools for emergent multilingual learners (9.4 and 9.8).

 To download the coaching tools for Chapter 6 that only have instructions for the teacher, go to **resources.corwin.com/mathematicscoaching.**

6.1 Connecting *Shifts* to Formative Assessment
Self-Assessment

Instructions to Coach: Ask teachers (individually or as part of a PLC activity) to self-assess where they position themselves on each of these Shifts in Classroom Practice *related to formative assessment. Use the following questions during a coaching conversation or in a PD setting to support teachers (and to help you decide which other tools may be most useful from this chapter). A version of this tool without this note is available for download.*

Instructions: The *Shifts in Classroom Practice* listed below have specific connections to formative assessment. Put an *X* on the continuum of each *Shift* to identify where you currently see your practice.

Tool 6.1 Shifts

Shift 1: From *stating-a-standard* toward *communicating expectations for learning*

Teacher shares broad performance goals and/or those provided in standards or curriculum documents.	Teacher creates lesson-specific learning goals and communicates these goals at critical times within the lesson to ensure students understand the lesson's purpose and what is expected of them.

Shift 5: From *questions that seek expected answers* toward *questions that illuminate and deepen student understanding*

Teacher poses closed and/or low-level questions, confirms correctness of responses, and provides little or no opportunity for students to explain their thinking.	Teacher poses questions that advance student thinking, deepen students' understanding, make the mathematics more visible, provide insights into student reasoning, and promote meaningful reflection.

Shift 8: From *looking at correct answers* toward *looking for students' thinking*

Teacher attends to whether an answer or procedure is (or is not) correct.	Teacher identifies specific strategies or representations that are important to notice; strategically uses observations, student responses to questions, and written work to determine what students understand; and uses these data to inform in-the-moment discourse and future lessons.

Tool 6.1 Reflection Questions

1. What do you notice, in general, about your self-assessment of these *Shifts in Classroom Practice*?

2. What might be specific teaching moves that align with where you placed yourself on the *Shifts*?

3. What might be specific teaching moves that align *to the right of* where you placed yourself on the *Shifts*?

4. What might be some professional learning opportunities to help you move to the right for one or more of these *Shifts*?

6.2 Planning for Five "Key Strategies" for Formative Assessment

Instructions: Use this tool to plan how you will implement the five key strategies (Wiliam, 2007) for effective formative assessment. Consider specific teacher practices within a lesson and within a unit or topic for each key strategy.

Key Strategies	Teacher Practices	
Clarifying, sharing, and understanding goals for learning and criteria for success with learners	Within a lesson	Within a unit or topic
Engineering effective classroom discussions, questions, activities, and tasks that elicit evidence of student learning	Within a lesson	Within a unit or topic
Providing feedback that moves learning forward	Within a lesson	Within a unit or topic
Activating students as owners of their own learning	Within a lesson	Within a unit or topic
Activating students as resources for one another	Within a lesson	Within a unit or topic

6.3 Developing Questions to Target Misconceptions

Instructions to Coach: Use this planning tool with teachers to anticipate common student misconceptions for each learning target in a standard or unit. Co-create questions that could be used during the lesson to support students in addressing misconceptions.

Instructions: Use this planning tool to anticipate common mathematical misconceptions for each learning target. Construct questions (for classroom discussions) to address the misconception.

Learning Targets/ Objectives	Anticipated Student Misconception(s)	Questions

6.4 Using Rubrics for Formative Assessment

Instructions to Coach: Use the rubric guide with teachers to identify a focus on learning targets and develop rubrics to use with students to self-assess and identify individual learning goals.

Instructions: Use this tool two ways. First, decide what you think performance will look like in each category related to your task/lesson (use the table below). Second, consider how you might use this tool with students to have them set performance levels for a project and/or self-assess.

Scoring With a Four-Point Rubric			
I can help others.	**I am there!**	**I am almost there.**	**I don't understand.**
I can explain my thinking using mathematical language so others understand. I can use several representations and make connections between mathematical ideas.	I can explain my thinking using mathematical language. I can show at least one representation. I know how what I am learning connects to other kinds of math.	I know what I am supposed to do, but I am a little confused about what I am learning. I know which representation makes sense for the math I am learning. I can make some connections.	I am confused. I do not know which representation to use, and I cannot make connections.

Lesson learning target(s):	
Task-Specific Language (What do students look like?)	
4: I can help others.	3: I am there!
2: I am almost there.	1: I don't understand.

Source: Adapted from The Formative 5: Everyday Assessment Techniques for Every Math Classroom *by Francis (Skip) Fennell, Beth McCord Kobett, and Jonathan A. Wray. Thousand Oaks, CA: Corwin, www.corwin.com. Copyright © 2017 by Corwin.*

6.5 Observing and Providing Feedback

Instructions to Coach: Work with teachers to use this tool during planning to anticipate how students will respond to a learning activity.

Instructions: Use this planning tool to connect your anticipated student responses to feedback you might provide to students.

Planning: Observations Template

Mathematics standard:			
Lesson objective:			
What would you expect to observe?	**How would you know "it" if you saw it?**	**What mathematical challenges or misconceptions might you observe?**	**How might you record and provide feedback of what you observe?**

Images: Clipart.com

 Available for download at **resources.corwin.com/mathematicscoaching.** Copyright © 2018 by Corwin.

6.6 Observing Students' Thinking

Instructions to Coach: Use this tool alongside teachers to collect evidence of students' understanding during a lesson. Then, unpack the observations together by sharing evidence and designing next steps.

Instructions: Collect evidence of student thinking by writing observations about student thinking. Record next steps that you want to take with the student.

Learning Target: _____

Level of Understanding	Anecdotal Evidence (Include Students' Names)	Next Steps
I can help others. • Has a clear understanding of concepts and related procedures • Can communicate concepts across multiple representations • Is able to illustrate understanding using various tools • Shows evidence of applying a strategy efficiently and effectively without prompting		
I am there! • Understands important concepts or procedures but makes minor errors • Is able to communicate concepts in some representations, but not across representations • Is able to illustrate understanding using various tools • May need prompting but can apply an effective strategy to demonstrate understanding		
I am almost there. • Demonstrates some understanding but also demonstrates confusion • Is able to perform an operation but cannot explain why it works or connect to a concept • With assistance, is able to illustrate a solution with at least one representation, but not across representations		
I don't understand. • Demonstrates difficulty in understanding the task • Cannot illustrate a solution with a representation • Is not able to perform an operation		

Source: Adapted from The Formative 5: Everyday Assessment Techniques for Every Math Classroom *by Francis (Skip) Fennell, Beth McCord Kobett, and Jonathan A. Wray. Thousand Oaks, CA: Corwin, www.corwin.com. Copyright © 2017 by Corwin.*

 Available for download at **resources.corwin.com/mathematicscoaching.** Copyright © 2018 by Corwin.

6.7 Observing Students' Representations

Instructions to Coach: Use this tool with teachers to plan for and anticipate students' use of representations in a lesson, collect data about those representations, and make decisions about the students who should share and provide feedback to peers.

Instructions: First, record all the representations you anticipate the students will use during your lesson. Then, as you teach, record using quick drawings or pictures of students' representations. Finally, select the students who will share representations with other students as a way to elicit and provide feedback to peers.

Classroom: Observation—Student Representations

Student Representations (Anticipated/Observed)	Who Is Using Specific Representations	Who I Will Select to Share Their Representations (Order of Presentations— 1st, 2nd, etc.)
Anticipated:		
Observed:		
Observed:		
Observed:		
Observed:		

Source: Adapted from Smith, M. S., and Stein, M. K. (2011). *Five Practices for Orchestrating Productive Mathematics Discussions.* Reston, VA: National Council of Teachers of Mathematics. Retrieved from the companion website for *The Formative 5: Everyday Assessment Techniques for Every Math Classroom* by Francis (Skip) Fennell, Beth McCord Kobett, and Jonathan A. Wray. Thousand Oaks, CA: Corwin, www.corwin.com. Reproduction authorized only for the local school site or nonprofit organization that purchased this publication.

 Available for download at **resources.corwin.com/mathematicscoaching.** Copyright © 2018 by Corwin.

6.8 Brief Formative Assessment Interview

Instructions to Coach: Ask teachers to conduct brief interviews during the lesson and bring the form to a coaching session. Work with teachers to unpack the information from the interviews and design the next instructional steps.

Instructions: Use this tool to record evidence of brief formative assessments you conduct in your lesson. Determine your next instructional steps using data gathered.

Individual Student: Interview Prompt

Student and Topic		
Name:	**Date:**	**Math topic:**

Question and Student Responses	Instructional Next Steps
1. How did you solve that?	
2. Why did you solve the problem that way?	
3. What else can you tell me about what you did?	

 Available for download at **resources.corwin.com/mathematicscoaching.** Copyright © 2018 by Corwin.

6.9 Five "Key Strategies" for Formative Assessment Data Collection

Instructions to Coach: Use this tool to record evidence of any of the five key strategies for effective formative assessment observed in the lesson. Keep in mind that every lesson may not include each strategy. This tool and Tool 6.2 can also be used in the reflecting conversation. Unpack your observations in a conversation with the teacher.

Instructions: Identify one or two key strategies (Wiliam, 2007) for your coach to "look for" while visiting your classroom. Decide what kinds of observable evidence you would like your coach to notice.

Key Strategy	Look For ...
1. Clarifying, sharing, and understanding goals for learning and criteria for success with learners **Look for ...** ○ Teacher shares learning goals with students. ○ Students discuss success in learning goals and/or criteria.	**Evidence:**
2. Engineering effective classroom discussions, questions, activities, and tasks that elicit evidence of student learning **Look for ...** ○ Teacher asks probing questions, prompts students to explain their thinking, and facilitates student discourse. ○ Students talk with one another about the mathematics they are learning. They justify their thinking and explain their reasoning.	
3. Providing feedback that moves learning forward **Look for ...** ○ Explicit feedback that directly connects to student understanding by encouraging students to think deeply about the mathematics. ○ Teachers provide timely and frequent feedback. ○ Students are able to explain and justify thinking.	
4. Activating students as owners of their own learning **Look for ...** ○ Teachers develop rubrics for students to self-assess learning that are directly tied to the learning targets. ○ Teachers connect learning activities to learning targets and success criteria and provide opportunities for students to reflect on their learning.	
5. Activating students as resources for one another **Look for ...** ○ Teachers use rubrics for students to peer-assess learning. ○ Teachers design learning activities and tasks for students to collaborate.	

6.10 Using Rubrics for Formative Assessment

Instructions to Coach: This tool can be used as a follow-up to Tool 6.4. *After the teacher has created and used the rubric, use this tool to reflect on the language and make revisions if necessary.*

Instructions: Use this tool to assess and revise language used in a rubric after having used it in order to make it more descriptive and useful to students (and yourself).

Lesson learning target(s):	
Task-Specific Language	
Original Language	*Revised Language (If Needed)*
4: Excellent	4: Excellent
3: Proficient	3: Proficient
3: Marginal	2: Marginal
1: Unsatisfactory	1: Unsatisfactory

- What aspects of the rubric seemed most helpful to students?
- In what ways did the rubric help you assess students' understanding of the learning target?
- What patterns are you noticing in students' understandings or misconceptions?

Previously published by Bay-Williams, J., McGatha, M., Kobett, B., and Wray, J. (2014). *Mathematics Coaching: Resources and Tools for Coaches and Leaders, K–12*. New York, NY: Pearson Education, Inc.

6.11 Analyzing Formative Assessment Key Strategies

Instructions to Coach: Encourage your teachers to reflect on the formative assessment process and practices in their classrooms by identifying successes and challenges as they relate to the key strategies. Help the teachers target a challenge and create steps for tackling the challenge.

Instructions: Reflect on your formative assessment practices in relation to these five key strategies (Wiliam, 2007) and identify successes and challenges. Select one challenge and identify specific steps for addressing the challenge.

Which of the five key strategies is most comfortable for you?	Which of the five key strategies is most challenging for you?
1. Clarifying, sharing, and understanding goals for learning and criteria for success with learners 2. Engineering effective classroom discussions, questions, activities, and tasks that elicit evidence of student learning 3. Providing feedback that moves learning forward 4. Activating students as owners of their own learning 5. Activating students as resources for one another	1. Clarifying, sharing, and understanding goals for learning and criteria for success with learners 2. Engineering effective classroom discussions, questions, activities, and tasks that elicit evidence of student learning 3. Providing feedback that moves learning forward 4. Activating students as owners of their own learning 5. Activating students as resources for one another
Describe:	Describe:
Using your success, how can you build your challenge? What can you do to build the key formative assessment strategy? Include specific action steps and a timeline.	

Chapter 7

Analyzing Student Work

Dear Coach,

Here is support for you as you work with teachers on analyzing student work.

In the COACH'S DIGEST ...

Overview: To review ideas about analyzing student work and/or download and share with teachers.

Coaching Considerations for Professional Learning: Ideas for how to support a teacher or a group of teachers in learning about analyzing student work.

Coaching Lessons From the Field: A mathematics coach shares her experience using a tool from this chapter.

Connecting to the Framework: Selected *Shifts in Classroom Practice* and Mathematical Practices are connected to analyzing student work.

Coaching Questions for Discussion: Menu of prompts for professional learning or one-on-one coaching.

Where to Learn More: Articles, books, and online resources for you and your teachers!

In the COACH'S TOOLKIT ...

Seven tools focused on analyzing student work, for professional learning or coaching cycles.

Coach's Digest

In the Coach's Digest, we begin with an overview of analyzing student work, written to teachers (and to you, the coach). As you read the Overview, the following questions might help you reflect on this topic in terms of your role as a coach:

- What kinds of shifts in teacher practices both inside and outside of the classroom can you support by developing learning communities to analyze student work?
- What types of professional learning activities, readings, and/or coaching cycles/lesson study might support teachers to develop a practice for regularly analyzing student work?
- How can you incorporate analyzing student work as a regular part of the support you provide teachers?

Overview of Analyzing Student Work

online resources ➚ | To download the Chapter 7 Overview, go to **resources.corwin.com/mathematicscoaching**.

Every day, teachers adjust and adapt instructional decisions in response to the work that they see students do in their classrooms. Analyzing student work is integral to planning, delivering instruction, and assessment. Understanding the complexity of student work and designing effective ways to analyze student **artifacts** is a powerful component of making important shifts in your classroom practice. This analysis can happen in two distinct ways: *inside* the classroom during the actual teaching or *outside* the classroom as teachers examine student work developed during the lesson.

The Purpose of Analyzing Student Work Inside the Classroom

Inside the classroom, teachers must quickly analyze student work to make instructional decisions. An example of this is when they facilitate learning and orchestrate discussions through meaningful tasks—they *anticipate* student responses to the task, *monitor* students' work as they solve the task, *select* and *sequence* some of the students' solutions to share, and then *connect* the students' solutions to learning goals (Smith & Stein, 2011). The important task of analyzing students' work happens during the *monitor, select,* and *sequence* stages of task facilitation. In this case, teachers analyze the students' solutions, looking for use of representations, connections to procedures, and understanding of mathematics concepts as they connect to the learning goal. The teacher's analysis of student work in this context is public and offers an opportunity to invite all students to join in the discussion with the goal of advancing student learning.

The Purpose of Analyzing Student Work Outside the Classroom

Teachers examine students' work to gain insights about what a student knows and to connect that insight to the way in which the topic was (or could be) taught. Analyzing student work can help improve instructional decision-making and target students' learning needs. While teachers' analysis of student work is often done in isolation (Little, Gearhart, Curry, & Kaftka, 2003), they experience benefits when the analysis is conducted in learning communities because they develop shared meaning about mathematics content and can connect particular kinds of student work to teaching practices (Kazemi & Franke, 2004). When teachers gather and reflect on student work, they offer and receive multiple interpretations of the work, which, in turn, invites sense making about students' thinking and deepens teachers' content knowledge (Colton & Langer, 2005).

This vision for student work requires substantial effort within a team or school. A review of schools successfully using student work to improve learning found three important schoolwide actions (Little et al., 2003):

- *Bring teachers together to focus on student learning and teaching practice.* When teachers brought evidence from their classrooms, they were able to make explicit connections about their students' work and the instructional practices that moved learning forward.
- *Get student work on the table and into the conversation.* Teachers looked at their own student work with more depth and analysis when it became part of the regular expected teacher conversations.
- *Structure the conversation.* Teachers developed protocols for sharing and analyzing student work.

Benefits of Analyzing Student Work

While the actions previously listed require an investment of time, this investment pays off! As teachers analyze student work, they make interesting and thoughtful connections between teaching practices and student learning. Both teachers and students benefit from increased attention to the work students create. Through purposeful analysis, they reap many benefits, including the following:

- *Intentional teaching.* Analyzing student work before a unit can yield excellent insights to use in designing lessons that build on students' strengths and address students' limited conceptions.
- *Deepened content knowledge.* Interpreting student strategies, alternative approaches, or errors may lead to insights and strengthen teachers' own content and pedagogical content knowledge.
- *A focus on Mathematical Practices.* Analyzing the proficiencies evident in student work helps clarify how students exhibit the Mathematical Practices. As you review student work, search for evidence of multiple representations, the selection of tools, the ability to provide a mathematical argument, and/or the exhibition of perseverance.
- *Efficient use of instructional time.* By taking time to analyze student work through prompts or problems, teachers can target their instruction to incorporate the identified strengths and weaknesses of the students.

Analyzing Student Work Protocol

Working with colleagues to analyze student work can be enhanced by using a professional learning protocol that invites inquiry, ignites purposeful instructional decision-making, and elicits reflection. Participating in conversations using the strategic norms from a protocol has numerous benefits, including opportunities to create shared understanding of content and teaching practices, to develop and test new ideas, and to establish goals for continuous improvement (Carr, Herman, & Harris, 2005; Crespo, 2002; Crockett, 2002; Little et al., 2003). Therefore, we offer a five-step collaborative student work analysis protocol for sharing and analyzing student work.

1. **Select student work.** The kind and quality of student work that is shared is important because the work must be meaty enough to engage teachers in discussions about students' mathematical understanding. In this step, participants collaboratively decide on a common task and the student work that will be collected from that task. If teaching different grade levels or courses, teachers can decide to collect student work that connects to a Mathematical Practice or content standards that progress across grade levels (i.e., vertical alignment). Sources for student work include written work from a task, brief interviews, or exit tasks. Using technology, student work can be gathered using interactive whiteboards (e.g., Show Me [www.showme.com] or Explain Everything [www.explaineverything.com]) or through an online formative assessment program (e.g., www.Goformative.com), which can be given to individual students or a whole class.

2. **Observe.** During this step, participants observe, without judgment, what they notice about the students' work. At this time, teachers refrain from jumping in to remedy misconceptions or offer evaluative feedback. They observe and record the students' strengths before identifying misconceptions. Analyzing the proficiencies evident in student work helps clarify the Mathematical Practices and supports teachers in being more intentional about developing these proficiencies in students. Teachers search for evidence of students' use of multiple representations as well as their tool selection, mathematical arguments, and/or

exhibited perseverance. All participants ensure that the observations are non-evaluative. For example, teachers analyzed the following student work:

$$\frac{2}{3} - \frac{1}{6} = \frac{1}{2}$$

I subtracted $\frac{1}{6}$

from $\frac{2}{3}$ and saw $\frac{1}{2}$

Strengths	Potential Misconceptions
• Student's procedures are correct. • Student included plan for solving. • Student drew a representation.	• Student's explanation does not show knowledge of equivalent fractions. • Student's representation does not match the procedures.

3. **Analyze and discuss.** In this step, questions are posed to offer explanations and interpretations of the student work. These ideas are also recorded before discussing.

> Analysis:
> Does the student remember how to do this this from doing the same problem?
> Is the student able to mentally subtract fractions?
> Did the student get lucky?
> Is more evidence needed to determine true understanding?

Once the analysis questions are offered, each idea is explored and discussed. At this time, the student's teacher clarifies and offers additional information about the student, classroom context, or task.

4. **Determine and implement next instructional steps.** After analyzing the work, teachers offer ideas for next instructional steps. This is a critical point in the protocol because participants are called to action as a result of deep conversations, uncovering student thinking by analyzing students' work. In this step, the focus shifts back to the teaching practices that will move learning forward. The discoveries about student thinking might encourage teachers to gather additional information (e.g., articles, books, resources), seek support from a coach, interview the student for additional information, and/or implement a new instructional task. For example, the analysis might have revealed misconceptions that require the teacher to move back to teaching at the concrete level.

5. **Share results.** In this step, teachers share the results of the student work analysis, including new insights into the student's understanding and adjustments to instruction. Teachers might share student work analysis results with school leadership as part of their documentation for effective teaching, with parents to communicate learning goals, or with other teachers to develop innovative teaching strategies.

Coaching Considerations for Professional Learning

Providing a structure for collaborative student work analysis is crucial because it establishes clear guidance on how to talk about the work in a way that is respectful and open. The overview provided a protocol for sharing and analyzing student work. In the following paragraphs are some ideas for organizing, collecting, and analyzing student work:

1. ***Engage teachers in developing or adjusting the protocol for sharing student work.*** Teachers sometimes feel nervous about sharing student work because they may feel protective of the student and/or insecure about their instructional approach. Protocols allow for nonjudgmental discussions and productive conversations that support deep analysis of student misconceptions. Some specific tools (see Tools 7.2, 7.6, and 7.7) are provided in this chapter and at the website Looking at Student Work (www.lasw.org/protocols.html).

2. ***Use different kinds of student work.*** The student work can be collected using video, problem-based tasks, photographs, web applications, student-constructed work, and/or journals. You can begin by asking teachers to bring in student work for discussion, such as a task they did in class. On other days, ask teachers to bring in video of students sharing their solutions. Using different forms of data provides different perspectives on student learning (see Tools 7.2 and 7.3). Focus discussions not only on what is learned from the data itself but also on the potential of that kind of data to provide insights on student learning (see Tools 3.4, 7.5, 7.6, and 7.7).

3. ***Vary the focus of the analysis.*** Sometimes an open-ended focus on what you notice from the work is valuable. But sometimes, as a coach, you want to adapt a protocol to focus on one particular aspect of teaching. You may want to identify (in advance or in the moment) a particular focus on conceptual understanding, procedural fluency (see Tools 3.3, 3.7, and 3.9), mathematical content (see Tool 3.4), or a Mathematical Practice (see Tools 2.4 and 2.9).

4. ***Target areas of challenge.*** Analyzing student work is intended to help teachers better meet the needs of students, so an effective way to do this is to ask teachers to bring in one piece of student work that is particularly problematic or confusing (see Tool 7.4). Teachers then use the protocol to analyze the student work objectively and make suggestions for instructional practice. This can be done as part of a PLC or part of a coaching cycle. As a way to focus this conversation, consider having teachers read any or all of the Overviews in Chapters 8, 9, and 10, which focus on meeting the needs of every student.

Coaching Lessons From the Field

I was both excited and terrified to conduct the *One Day/Many Artifacts* event with my school. I had been preparing them to be thinking about the kinds of artifacts they wanted to bring. I used the planning sheet to help me focus my ideas before I presented them to the staff.

We decided to focus on the Mathematics Practices, which turned out to be a huge hit! The teachers were really proud to share their students' work. I think the event was a strength-based, community-building event. My principal has asked me to do this once a semester!

One Day/Many Artifacts Planning

1. What types of student work might be collected (pictures, video, pieces of student work)?

Math Practices - Will encourage teachers to find and bring artifacts that show students engaging in the practices. Would love to see them bring lots of different types of artifacts. Need to talk to Fred's team. They have been using Goformative – will see if they can share some work. Also, check with Jen's team. They have been using the Show Me app.

2. What might the expectations be for teachers to contribute student work?

I expect that every teacher will bring at least one artifact. May need back up from Darius.

3. How might the work be collected?

I will ask them to let me know ahead of time what they plan to bring and then prepare table tents for them to display. Maybe I can collect the artifacts in the team planning meetings?

4. How might the work be stored? Sorted?

Stored in my office by team. I will have the teachers sort them by what they see. Maybe colored dots with the practices?

Connecting to the Leading for Mathematical Proficiency (LMP) Framework

As teachers focus on analyzing student work, it is important for them to make explicit connections to the *Shifts in Classroom Practice* and the Mathematical Practices. The brief paragraphs that follow provide ideas for making these connections. You and your teachers can continue to add to these ideas (see Tool 7.1 for connecting to the *Shifts*).

Connecting Analyzing Student Work to *Shifts in Classroom Practice*

Shifts	1	2	3	4	5	6	7	8

Shift 1: From *stating-a-standard* toward *communicating expectations for learning*

Teacher shares broad performance goals and/or those provided in standards or curriculum documents.	Teacher creates lesson-specific learning goals and communicates these goals at critical times within the lesson to ensure students understand the lesson's purpose and what is expected of them.

As teachers consider the work they will analyze, they design or select tasks that will reveal students' mathematical thinking as it connects to the learning goal. Teachers design opportunities throughout the lesson to collect student work products.

Shift 4: From *show-and-tell* toward *share-and-compare*

Teacher has students share their answers.	→ Teacher creates a dynamic forum where students share, listen, honor, and critique each other's ideas to clarify and deepen mathematical understandings and language; teacher strategically invites participation in ways that facilitate mathematical connections.

As teachers facilitate learning tasks, they monitor students' work and encourage them to share solution strategies. Encourage teachers to analyze student work and then use that work to facilitate other students' understanding. Offer opportunities for teachers to share how they select and sequence students' work to share with classmates.

Shift 5: From *questions that seek expected answers* toward *questions that illuminate and deepen student understanding*

Teacher poses closed and/or low-level questions, confirms correctness of responses, and provides little or no opportunity for students to explain their thinking.	→ Teacher poses questions that advance student thinking, deepen students' understanding, make the mathematics more visible, provide insights into student reasoning, and promote meaningful reflection.

Sometimes the students' solutions are unexpected. Rather than focus on the students, it can be helpful to examine the task or prompt that elicited that particular student work. As teachers design tasks, they need to consider how the task will invite students to demonstrate their thinking. Teachers who anticipate how students will respond to particular tasks can compare expected responses to actual student responses.

Shift 8: From *looking at correct answers* toward *looking for students' thinking*

Teacher attends to whether an answer or procedure is (or is not) correct.	→ Teacher identifies specific strategies or representations that are important to notice; strategically uses observations, student responses to questions, and written work to determine what students understand; and uses these data to inform in-the-moment discourse and future lessons.

Analyzing students' work requires that teachers examine solutions with a mindset for understanding student thinking rather than looking for correctness of responses. Encourage teachers to unpack students' strategies by examining mathematical representations, strategies, and explanations. Encourage teachers to use evidence from students' work to adjust or adapt instruction.

> → Ask your teachers to connect analyzing student work to the *Shifts* by having them think about how understanding students' thinking can inform instructional decision-making. Ask your teachers to identify particular student misconceptions first. Then, have them determine a *Shift* that will help them identify the next instructional steps to move students' mathematical understanding forward.

Connecting Analyzing Student Work to Mathematical Practices

MPs   **1**   2   **3**   4   **5**   **6**   7   8

An important component of analyzing student work is looking for evidence of the Mathematical Practices. Depending on the type of student work, different Mathematical Practices can be demonstrated by students and ultimately could be supported by all the Practices. Mathematical Practices 1, 3, 5, and 6 are naturally connected to analyzing student work because the student product or artifact provides evidence of student thinking.

1. ***Make sense of problems and persevere in solving them.*** When analyzing student work, teachers can look to see whether there is evidence of students selecting an effective strategy and checking their work. Students do this by solving the problem in different ways. They may also indicate this by finding an incorrect answer and changing directions within the problem. If the artifact is developed through **pencast** or video technology, student perseverance is demonstrated when students stick with a problem, reason mathematically, ask questions out loud, change directions in their thinking, and persevere through problems.

3. ***Model with mathematics.*** When students create or connect real-world models for the mathematics they are learning to demonstrate their understanding of a situation; they are making connections. Further examination of the models reveals explicitly whether students hold specific misconceptions about the mathematics concepts or procedures.

5. ***Use appropriate tools strategically.*** Mathematically proficient students can use appropriate tools (calculators, manipulatives, graph paper, etc.) to solve mathematical tasks. Analyzing the choice of tools and methods students use can reveal interesting patterns in student thinking. Students with misconceptions often use tools without considering their efficiency or appropriateness. Students with strong conceptual understanding explicitly link their tools with the mathematics.

6. ***Attend to precision.*** Procedural fluency can be analyzed within student work as teachers examine procedures, methods, and algorithms students use to solve problems. Teachers can analyze student work to see whether students are learning precise ways of describing mathematical ideas. Providing feedback to students and requesting revisions are effective teaching practices.

→ Ask your teachers to connect the Mathematical Practices to analyzing student work by asking the following questions:

- What connections do you see between analyzing student work and the selected Mathematical Practices?

- What adjustments have you made as a result of analyzing student work that support students in demonstrating the Mathematical Practices?

- How does planning for collecting student work support students' use of the Mathematical Practices?

Coaching Questions for Discussion

Questions Related to the Focus Zone: Analyzing Student Work

1. In what ways does a focus on analyzing student work help teachers connect evidence from student work to teaching practices?

2. How might engaging in deep analysis of student work improve your planning and instruction?

3. How might you use evidence from student work to design rich lessons?

4. What might student work tell you about teachers' instructional practices?

5. How might you collect student work as a teacher, PLC team, or school?

Questions Related to the LMP Framework

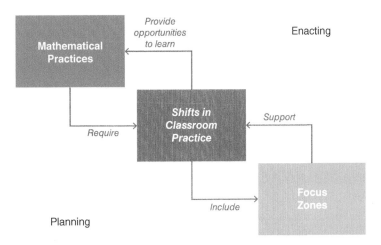

1. As you enact the ideas of analyzing student work, what do you see as related *Shifts in Classroom Practice*?

2. As you enact the ideas of analyzing student work, what do you see as opportunities to learn and, in particular, to develop the Mathematical Practices?

Where to Learn More

Books

Ashlock, R. (2006). *Error Patterns in Computation: Using Error Patterns to Improve Instruction* (9th ed.). Boston, MA: Pearson.

> *Ashlock provides a model for examining student work in order to determine typical error patterns. The text also considers underlying reasons for these errors and offers strategies for student intervention. The focus is on computation, but the strategies for analyses can be applied to other subject areas.*

Goldsmith, L. T., & Seago, N. M. (2013). *Examining Mathematics Practice Through Classroom Artifacts.* Boston, MA: Pearson.

> *This book supports teachers in learning how to skillfully use classroom artifacts to analyze student work to gain insights into students' mathematical thinking. A PD Toolkit, including a website with videos and worksheets, accompanies the book.*

Tapper, J. (2014). *Solving for Why: Understanding, Assessing, and Teaching Students Who Struggle With Math.* Sausalito, CA: Math Solutions.

> *This book explores the reasons underlying students' mathematical struggles and includes tools, approaches, and suggestions for implementing interviews.*

Online Resources

Beliefs About Student Work

www.sdcoe.net/lret2/dsi/pdf/LASW_Beliefs-Protocol.pdf
> *Developed by the San Diego County Office of Education, this resource is a guide for conducting collaborative data conversations around student work.*

GoFormative

www.goformative.com
> *This is a web-based application for collecting student work.*

Using Student Achievement Data to Support Instructional Decision-Making (Doing What Works)

http://dww.ed.gov
> *Based on the IES practice guide by Hamilton et al. (2009), this resource provides coaches and teacher leaders with tools and resources to support successful data decision-making.*

Coach's Toolkit

These tools are a menu from which you can select any that make sense for your setting/context. They can be used independently or as part of a coaching cycle. You may start with the self-assessment, which can guide you in deciding which of the other tools may be most useful. If using these tools for a coaching cycle, mix and match as you like or use one of the combinations we suggest in the diagrams that follow. The tools in this chapter include instructions to the coach and the teacher. You can download copies of the tools that only have instructions for the teacher at **resources.corwin.com/ mathematicscoaching**.

Self-Assess

7.1	Connecting *Shifts* to Analyzing Student Work Self-Assessment

Gather Data

7.4	Understanding Student Thinking
7.5	Analyzing One Student's Work

Plan

7.2	One Day/Many Artifacts Planning
7.3	Planning Task Implementation

Reflect

7.6	Analysis of Students' Misconceptions
7.7	Collaborative Analysis Protocol

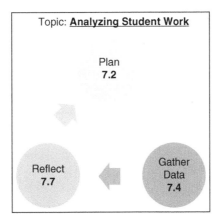

Additional Tools From Other Chapters

Tools in Chapter 6 (Formative Assessment) can also be used to support analyzing student work; Tools 6.3, 6.6, and 6.7 specifically target student thinking. Tools in Chapters 9 include diagnostic interview tools for emergent multilingual learners (9.4 and 9.8). Tools in Chapters 2 and 3 focus on developing students' content knowledge, procedural fluency, and use of Mathematics Practices, which can serve as a focus for the analysis.

 To download the coaching tools for Chapter 7 that only have instructions for the teacher, go to **resources.corwin.com/mathematicscoaching**.

7.1 Connecting *Shifts* to Analyzing Student Work
Self-Assessment

Instructions to Coach: Ask teachers (individually or as part of a PLC activity) to self-assess where they position themselves on each of these Shifts in Classroom Practice *related to analyzing student work. Use the following questions during a coaching conversation or in a PD setting to support teachers (and to help you decide which tools may be most useful from this chapter). A one-page version of this tool without this note is available online for download.*

Instructions: The *Shifts in Classroom Practice* listed below have specific connections to engaging the learner. Put an *X* on the continuum of each *Shift* to identify where you currently see your practice.

Tool 7.1 Shifts

Shift 1: From *stating-a-standard* toward *communicating expectations for learning*

Teacher shares broad performance goals and/or those provided in standards or curriculum documents. ⟶ Teacher creates lesson-specific learning goals and communicates these goals at critical times within the lesson to ensure students understand the lesson's purpose and what is expected of them.

Shift 4: From *show-and-tell* toward *share-and-compare*

Teacher has students share their answers. ⟶ Teacher creates a dynamic forum where students share, listen, honor, and critique each other's ideas to clarify and deepen mathematical understandings and language; teacher strategically invites participation in ways that facilitate mathematical connections.

Shift 5: From *questions that seek expected answers* toward *questions that illuminate and deepen student understanding*

Teacher poses closed and/or low-level questions, confirms correctness of responses, and provides little or no opportunity for students to explain their thinking. ⟶ Teacher poses questions that advance student thinking, deepen students' understanding, make the mathematics more visible, provide insights into student reasoning, and promote meaningful reflection.

Shift 8: From *looking at correct answers* toward *looking for students' thinking*

Teacher attends to whether an answer or procedure is (or is not) correct. ⟶ Teacher identifies specific strategies or representations that are important to notice; strategically uses observations, student responses to questions, and written work to determine what students understand; and uses these data to inform in-the-moment discourse and future lessons.

Tool 7.1 Reflection Questions

1. What do you notice, in general, about your self-assessment of these *Shifts in Classroom Practice*?

2. What might be specific teaching moves that align with where you placed yourself on the *Shifts*?

3. What might be specific teaching moves that align *to the right* of where you placed yourself on the *Shifts*?

4. What might be some professional learning opportunities to help you move to the right for one or more of these *Shifts*?

7.2 One Day/Many Artifacts Planning

Instructions to the Coach: Use this form to help teams decide on artifacts. Collect student work from the entire school or a team of teachers on one day. Artifacts can be collected through interviews, pictures, papers, video, and web applications. Ask teachers to analyze artifacts to determine themes or trends within one team or in cross-grade-level meetings. Discuss implications for teaching.

1. What types of student work might be collected (pictures, video, pieces of student work)?

2. What might the expectations be for what student work each teacher will contribute?

3. How might the work be collected?

4. How might the work be shared, stored, and sorted?

Tool previously published by Bay-Williams, J., McGatha, M., Kobett, B., and Wray, J. (2014). Mathematics Coaching: Resources and Tools for Coaches and Leaders, K–12. New York, NY: Pearson Education, Inc.

7.3 Planning Task Implementation

Instructions: Use this tool first to anticipate students' thinking. Then, use the tool as a data collection tool to record student work as it connects to teacher questions, moves, and practices.

Task

Launching the Task	
Teacher Statements/Questions	Student Statements/Questions

Facilitating the Task	
Teacher Statements/Questions	Student Statements/Questions

Sharing/Presenting Task Solutions	
Teacher Statements/Questions	Student Statements/Questions

7.4 Understanding Student Thinking

Instructions to the Coach: Use this tool as a PLC or for a lesson cycle. If part of a lesson cycle, this tool can be given to teachers to gather data on a student. Or both you and the teacher can gather data on a student and then compare your data collection.

Instructions: Observe a student solving a problem/task. Ask probing or clarifying questions to determine student thinking.

Description of task observed

Observations of and explanations by student

1. What prior knowledge did the student have to have in order to solve this problem?	2. What strategies did the student use to solve the problem? Did the student try different approaches?
3. In what ways did the questions you asked extend the student's thinking?	**4. What was learned from the task itself and/or the questions you posed?**

Tool previously published by Bay-Williams, J., McGatha, M., Kobett, B., and Wray, J. (2014). Mathematics Coaching: Resources and Tools for Coaches and Leaders, K–12. *New York, NY: Pearson Education, Inc.*

7.5 Analyzing One Student's Work

Instructions: Use this tool to record your analysis of a student work sample.

Student name/Grade level:

Description of student's work (How and when was the student work collected?)

Student's strengths	Student's misconceptions

Future work with the student

Tool previously published by Bay-Williams, J., McGatha, M., Kobett, B., and Wray, J. (2014). Mathematics Coaching: Resources and Tools for Coaches and Leaders, K–12. *New York, NY: Pearson Education, Inc.*

7.6 Analysis of Students' Misconceptions

Instructions: Sort the students' work with a focus on errors and misconceptions. After sorting, complete this template. Determine next steps for the student(s).

Common Student Misconceptions	Uncommon Student Misconceptions

Next Steps Instructional Moves	Next Steps Instructional Moves

7.7 Collaborative Analysis Protocol

Instructions: As a team, decide on a content or Mathematical Practice focus for the collaborative analysis. Bring samples of student work to analyze in a PLC that reflect the focus.

Student	Task

Observations	Strengths
	Misconceptions

Analyze and discuss (pose questions)

Determine and implement next instruction steps

Chapter 8

Differentiating Instruction for All Learners

Dear Coach,

Here is support for you as you work with teachers on differentiating instruction for all learners.

In the COACH'S DIGEST ...

Overview: Highlights about differentiating instruction for you and/or to download and share with teachers.

Coaching Considerations for Professional Learning: Ideas for how to support a teacher or a group of teachers in learning about differentiated instruction.

Coaching Lessons From the Field: Story from a mathematics coach who has worked with teachers on differentiating instruction.

Connecting to the Framework: Specific ways to connect selected *Shifts* and Mathematical Practices to differentiating instruction.

Coaching Questions for Discussion: Menu of prompts for professional learning or one-on-one coaching about differentiated instruction.

Where to Learn More: Articles, books, and online resources for you and your teachers to learn more about this topic!

In the COACH'S TOOLKIT ...

Nine tools focused on differentiated instruction, for professional learning or coaching cycles.

Coach's Digest

In the Coach's Digest, we begin with an overview of **differentiating instruction**, written to teachers (and to you). As you read the Overview, the following questions might help you reflect on this topic in terms of your role as a mathematics coach:

- How might I help teachers figure out efficient ways to implement differentiated instruction?
- Which of these strategies or ideas might be a starting place for my teachers (or shall I ask them to select one)?
- How might we integrate practices such as opening tasks into our regular planning routines?

Overview of Differentiating Instruction

online resources ↖ To download the Chapter 8 Overview, go to **resources.corwin.com/mathematicscoaching.**

Differentiation is the act of including strategies to support the range of different academic backgrounds in classrooms (Tomlinson, 2001). The Mathematical Practices and the *Shifts in Classroom Practice* essentially describe a differentiated classroom—one in which students can select strategies to solve problems, demonstrate understanding in a variety of ways, and so on. Such a problem-based approach to teaching attends to the range of learners, while a teacher-demonstration approach to teaching tends to treat all learners the same.

Differentiating Instruction: The Basics

Differentiating instruction means thinking of the students as a set of individuals, each with his or her own strengths and learning needs, and trying to make sure that the lesson is designed in such a way that each student has access to the content. This includes accommodations and modifications for students with special needs, gifted students, students who are ethnically and culturally diverse, **emergent multilingual students**, and students who are struggling. (Note that we use the term *emergent multilingual students* [discussed more in Chapter 9] to describe students who are learning English as they are learning content. The term replaces the more commonly used phrase *English Language Learners [ELLs]* because it is a more accurate and inclusive description of these students.) Differentiating instruction is based on these three essential elements (Small, 2017; Smith, 2017a; Smith 2017b; Sousa & Tomlinson, 2011):

1. *Focusing on meaningful content, including authentic contexts.* An **authentic context** is a situation or story that is familiar to the lived experiences of a student and that is an actual context in which the selected mathematics would be used. Content must be developmentally appropriate and emphasize the Mathematical Practices (see Chapters 1 and 2).

2. *Recognizing each student's readiness, interests, and learning preferences.* These are the resources students bring to learning. In planning, we consider these questions:
   - *Readiness:* What do students know related to the topic, including the topic itself and any prerequisite knowledge?
   - *Interest:* What books, activities, and hobbies do students enjoy, and what personal, school, and community issues might they want to explore mathematically?
   - *Learning preferences:* In what ways does a student like to work (independently, in groups), and what modes support her or his thinking (visual, concrete, auditory, etc.)?

3. ***Connecting learner to content, including choices in content, process, and products.*** Differentiation in a lesson includes providing options for students in any or all of three domains—content, process, and product. In planning, we consider these questions:
   - *Content:* How might I modify or adapt the content of a lesson? *Modify* means changing the actual lesson or task. For example, a task of ordering fractions, decimals, and percentages may be modified so that some learners are ordering fractions only. Modifications must not be simplifications! Simplifying is not going to lead to increased learning. *Accommodations* leave the task unaltered, but change how it is enacted in order to ensure that the task is accessible and effective. Accommodating the lesson on fractions might include using manipulatives or representations, providing non-examples and examples, and increasing attention to mathematics language (Cassone, 2009).

- *Process:* How might students engage with the content? Tomlinson (1999) described this as taking "different roads to the same destination" (p. 12). In giving students choice, the process connects to students' readiness, interests, and learning preferences (Tomlimson & McTighe, 2006). Students, for example, might select representations, solution strategies, or tasks.
- *Product:* What will students show, write, or tell to demonstrate what they have learned? The products for a single task are the ways students share their ideas at the end of a lesson, which might include students showing (e.g., with graphing technology or manipulatives), writing, or telling. The products related to the unit or project might offer choices, such as presentations, written projects, or a test.

4. ***Creating differentiated tasks in manageable ways.*** Descriptions of differentiated instruction can sound like three times the work, creating three lessons for one class. This is not reasonable, given the lack of planning time in US classrooms! It is important to have efficient strategies for differentiating instruction, such as open questions, **tiered tasks**, and **parallel tasks**. *Open questions* are broad-based questions that invite meaningful responses from students at many developmental levels (Small, 2017). A question is open when it is worded in such a way that a variety of approaches or answers are possible (see also the worthwhile tasks discussion in Chapter 2). For example,
   - Rather than simply directing *Find the area of the rectangle*, ask *What options are possible for a rectangle with an area of 24 square units?*
   - Rather than instructing *Write _____ in scientific notation*, ask *What two numbers are easier to add when in scientific notation than in standard form?*

This general strategy is to use the "answer" and seek a possible "problem" to go with it. Another quick strategy is to ask students to describe similarities and/or differences between two problems; this has the added benefit of moving from a low-level task to a high-level task (Small, 2017).

Tiered lessons have a series of options for students, and those options might vary in terms of the structure, complexity of the task, complexity of the process, and degree of assistance (Kingore, 2006). Figure 8.1 provides an example that was created in about 15 minutes. The point is that it doesn't have to be perfect or fancy, but opening this task and varying its structure has resulted in a lesson that will teach the content more deeply and be accessible to all students.

Parallel tasks, like tiered lessons, involve students working on different tasks all focused on the same learning goal, but with parallel tasks, the focus is on *choice*. Consider this example, based on Small and Lin (2010):

Find the equation of the line that completes the polygon.

Option 1: A parallelogram with three sides formed by $y = 2$, $y = 5$, and $y = -2x + 1$

Option 2: A right triangle with two sides formed by $y = -2x + 8$ and $y = \frac{1}{3}x$

There are often two textbook exercises that can be selected to become parallel tasks (e.g., solve either #5 or #8). The key is that the two tasks both address the same mathematical goals and that questions can be discussed that apply to both. In this case, for example, the teacher might ask *Is there more than one possible line?*

Figure 8.1 **Example of Textbook Task Modified Into Tiered Lesson**

Textbook task: For each shape, tell how many angles and how many vertices (five polygons are illustrated, including a triangle, parallelogram, square, pentagon, and hexagon).

Observation: This is closed and is lower level.

Goal: Open it and tier how much structure is provided.

Redesigned tiered task:

Less structured: Explore your shapes and write three things you notice about their sides and their angles. Use words and pictures to illustrate what you notice.

Medium structured: Think of a rule you can use to sort your shapes. Record the rule. Draw a picture of a shape that fits the rule and a shape that does not fit the rule.

Highly structured: Sort your shapes using the table. After you have completed the table, describe patterns you notice about the sides and angles.

Shapes	Number of Sides	Number of Angles
	3	
		4
	5	
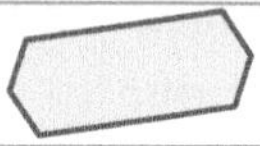		

NOTES

_____

_____

_____

_____

_____

_____

_____

_____

_____

_____

_____

_____

_____

_____

_____

_____

_____

_____

_____

_____

_____

Coaching Considerations for Professional Learning

Meeting the needs of all learners can appear to require far more planning time than teachers have, particularly new teachers. Focusing on how to take specific steps to differentiate can help teachers view differentiation as accessible (and necessary). Here are some ideas for helping teachers increase their skill at differentiating instruction:

1. *Make "high expectations" a tangible term.* The term *high expectations* may be known and even overused, but so is the practice of lowering the level of cognitive demand and avoiding high-level thinking tasks when differentiating to support students who struggle. It is therefore very important to reflect on any modification or adaptation in terms of (a) the way in which it increases access (and for whom) and (b) to what extent the mathematical challenge is still intact. Providing examples and non-examples for teachers of tasks that are modified but not simplified and then having them generate tasks themselves can be a powerful professional learning experience. Tools 8.2, 8.3, and 8.4 have a focus on determining strategies or accommodations for students, and these can be discussed in terms of keeping expectations high.

2. *Address how to "help."* Perhaps one of the greatest pitfalls in teaching is that in the interest of "helping" a student, the thinking is taken away. Adding structure to a task can help; showing students how to solve it does not. These are subtle nuances that are part of the enacting of the differentiated lesson. Therefore, videos, vignettes, or lesson observations are needed in order to reflect on whether the helping is getting the students to an answer or supporting their journey toward mathematical proficiency (see Tools 8.6 and 8.7, which can be used with videos or with classroom observations).

3. *Help locate differentiated tasks and good resources.* Tasks that lend themselves to multiple strategies and incorporate high-level thinking (open tasks) are more accessible to a range of learners. Yet finding these tasks on the Internet is not as easy as finding low-level worksheets. Knowing where to find such tasks or having structured time to explore and discuss such tasks will help (see Where to Learn More later in this chapter).

4. *Engage teachers in adapting and modifying tasks.* Because teachers daily encounter closed and low-level-thinking mathematics tasks and don't have a lot of time to look for more variety in tasks, being good at adapting tasks can greatly improve students' opportunities to learn in their classrooms (see Tools 8.4 and 8.5). The more practice teachers have in adapting tasks, the quicker and better they get at doing it. Therefore, it can become a regular focus for professional learning gatherings and/or coaching cycles. For example, in a workshop setting teachers might work in breakout groups to adapt a task using any of the strategies described in the Overview. Tool 8.10 can be used as a follow-up to reflect on the impact of using an adapted task.

5. *Modify tasks to incorporate Mathematical Practices.* These critical proficiencies take a back seat to content demands if we are not intentional about incorporating them. Teachers might explore a task, as described in #4, and then discuss which Mathematical Practices are supported in the revised task. Don't stop there! Just naming possible matches is not enough. An important next step is to ensure that the selected Mathematical Practice is prominent in the lesson (which may lead to further modifications or adaptations). Another way to approach this activity is to have teachers solve a task (e.g., from their textbook or favorite resource) and then be assigned a Mathematical Practice and asked *How can you modify or adapt this task in order to provide opportunities for your students to develop Mathematical Practice #___?* Doing so for one task will give teachers ideas of how to emphasize this practice on other tasks and thereby be more effective in developing mathematical proficiency.

Coaching Lessons From the Field

Helping my teachers open up closed tasks is so important to increasing students' opportunities to do the Math Practices! And so much of what they have available to them in their textbooks or from the Internet is closed tasks (and low level). I have teachers read an article about opening closed tasks (I like "Turning Traditional Textbook Problems Into Open-Ended Problems"—see Articles). We have used this article in a professional learning session with the Jigsaw strategy. Each expert group gets one of the content domains (no one has the full article), and everyone reads the introduction. When they return to their home group, they ask their peers how *they think it was opened* and then share how it was opened. Next, I give them a chance to practice on closed tasks from their textbook (I have also brought ones from textbooks). Finally, they prepare one open task that they will implement, keep the student work, and bring it with them to the next session.

Connecting to the Leading for Mathematical Proficiency (LMP) Framework

As teachers focus on differentiating instruction, it is important for them to make explicit connections to the *Shifts in Classroom Practice* and the Mathematical Practices. The brief paragraphs that follow provide ideas for making these connections. You and your teachers can continue to add to these ideas (see also Tool 8.1 for connecting to the *Shifts*).

Connecting Differentiating Instruction to *Shifts in Classroom Practice*

Shifts	1	2	3	4	5	6	7	8

Differentiating instruction relates to every *Shift* and is critical to ensuring the success of every student. Here, we briefly share connections for four of the *Shifts*.

Shift 1: From *stating-a-standard* toward *communicating expectations for learning*

Teacher shares broad performance goals and/or those provided in standards or curriculum documents. $\longrightarrow$ Teacher creates lesson-specific learning goals and communicates these goals at critical times within the lesson to ensure students understand the lesson's purpose and what is expected of them.

A strong lesson goal or objective, while clear and specific, does not limit the ways in which that content can be demonstrated. To really differentiate in a way that supports all learners, the content, process, and product options must each result in meeting the purpose of the lesson. When one option of a tiered lesson involves only low-level thinking while another involves high-level thinking, the tasks are not focused on the same learning goals.

Shift 3: From *teaching about representations* toward *teaching through representations*

Teacher shows students how to create a representation (e.g., a graph or picture). $\longrightarrow$ Teacher uses lesson goals to determine whether to highlight particular representations or to have students select a representation; in both cases, teacher provides opportunities for students to compare different representations and how they connect to key mathematical concepts.

Differentiating lessons makes this *Shift* possible. Students can only compare representations when they have been assigned different representations to produce (e.g., from a tiered task) or if they have had a choice of representations (e.g., through an open or parallel task or how instructions are framed). Tiered tasks or parallel tasks can be employed by the teacher to highlight important mathematical concepts. It bears repeating: Differentiated or open tasks are not only to ensure students have access to the mathematics; they are also to ensure students have the opportunity to compare and contrast strategies, raising the level of thinking in the lesson.

Shift 7: From *mathematics-made-easy* toward *mathematics-takes-time*

Teacher presents mathematics in small chunks so that students reach solutions quickly. → Teacher questions, encourages, provides time, and explicitly states the value of grappling with mathematical tasks, making multiple attempts, and learning from mistakes.

Opening up a task (and raising its level of cognitive demand) is communicating to students that doing mathematics takes time. Opening a task, along with encouraging different solution pathways, tells students that there actually are multiple pathways and should one of them not be working for them, they can try another pathway. Finally, choice is one of the most effective ways to motivate students, and that motivation can help students stick with a task.

Shift 8: From *looking at correct answers* toward *looking for students' thinking*

Teacher attends to whether an answer or procedure is (or is not) correct. → Teacher identifies specific strategies or representations that are important to notice; strategically uses observations, student responses to questions, and written work to determine what students understand; and uses these data to inform in-the-moment discourse and future lessons.

When focusing on differentiation, this *Shift* is at the center of that work (and vice versa). Once tasks are opened and students are solving them in a variety of ways, the focus cannot be on whether one procedure was done correctly; it must be on whether the student's explanation, representation, or process meets the learning goals of the lesson.

> → One way to help teachers connect differentiated instruction to the *Shifts* is to give them a differentiated task (e.g., a parallel task) and have them think about how they would engage students with the task before they solve it, as they are working on it, and after they have completed it. Then, map those ideas across one or all of the *Shift* continua (see Tool 8.1). A powerful contrast is to do this process with a closed task (e.g., from their textbook), having them consider what they might say/do when using that task or worksheet, and look at where those actions fall on the continua. Next, have teachers open the task (see Tool 8.5), consider how they would implement it, and map those ideas to the *Shifts*. Hopefully, they will see that differentiated tasks provide increased opportunities for effective teaching (and thereby better learning).

Connect Differentiating Instruction to Mathematical Practices

MPs	1	2	3	4	5	6	7	8

Meeting the needs of all learners permeates all of the Mathematical Practices! But in order for us to "dig deep" into differentiated instruction, we describe three that have explicit and critical connections.

1. ***Make sense of problems and persevere in solving them.*** At the heart of this practice is valuing multiple solution routes (based on prior knowledge and experiences), and similarly, that is the focus of differentiated instruction. A good task has multiple entry points—meaning that if one student does not know one way to solve it, he or she can solve it a different way. Secondly, building perseverance matters. The longer a student tries to solve a problem, the more likely he or she is to meet with success! How a task is structured can impact the extent to which a student is willing or able to persevere.

3. ***Construct viable arguments and critique the reasoning of others.*** If all students solve the same problem or if all students solve problems the same way, there is no reason to prepare a mathematical argument and no opportunity to critique the reasoning of others. When students have different ways to solve problems through open, tiered, or parallel tasks, they have an opportunity to create mathematical arguments and evaluate other mathematical arguments (both activities are at the highest cognitive levels). Thus, differentiating instruction is not just important for meeting the needs of each learner, but for collectively raising the level of thinking in the classroom.

5. ***Use appropriate tools strategically.*** The focus of this Mathematical Practice is on *students* choosing what tools are needed to solve a problem. One student may choose a manipulative, another grid paper, and another an Excel spreadsheet. Having choice naturally differentiates the process by which students engage in the lesson and the product they will share. And choosing tools means that students are selecting a strategy that makes sense to them, making the mathematics more accessible to all students.

Ask your teachers to connect the Mathematical Practices to differentiating instruction by asking these questions:

- What connections do you see between differentiation and the selected Mathematical Practices?

- In what ways have you differentiated instruction that have been successful in supporting each student's use of Mathematical Practices? (And which Mathematical Practices do you notice?)

- As you have implemented differentiated tasks (open, tiered, or parallel), what have you noticed in terms of students making sense of the mathematics, critiquing reasoning, and using tools strategically?

Coaching Questions for Discussion

Questions Related to the Focus Zone: Differentiated Instruction

1. Who are the various students in your classroom? What challenges and resources does each student bring?

2. Mathematical Practice 1 encourages using tasks with multiple entry points. What good examples have you used? In what ways might a task/lesson be modified to ensure it has multiple entry points?

3. What strategies might be needed to ensure all students are engaged in a lesson? What ideas do you have for adding structure to a lesson but retaining high-level thinking?

4. How have you, and how might you, communicate support for using students' own strategies in doing mathematics?

5. To what extent have you explored differentiating instruction? In which of the ways have you focused your differentiation efforts (content, process, product)?

6. As you implement a differentiated task, what talk moves or engagement techniques might you incorporate into each phase of your lesson (in introducing the task, as students work on the task, during whole class debriefing of the task)?

7. As students work on different tasks or strategies, how will they support each other?

8. How will you engage students in learning about each other's tasks and strategies?

Questions Related to the LMP Framework

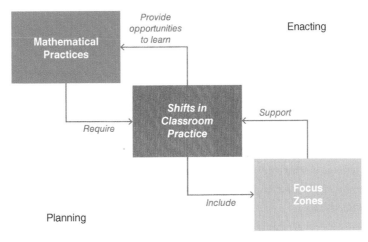

1. As you enact the ideas of differentiating instruction, what do you see as related *Shifts in Classroom Practice*?

2. As you enact the ideas of differentiating instruction, what do you see as opportunities to learn and, in particular, to develop the Mathematical Practices?

Where to Learn More

Books

Aguirre, J. M., Mayfield-Ingram, K., & Martin, D. B. (2013). *The Impact of Identity in K–8 Mathematics: Rethinking Equity-Based Practices.* Reston, VA: NCTM.

Drawing upon students' identities is critical to offering high-quality mathematics instruction! This book describes five practices for equitable mathematics teaching, with differentiation themes embedded. And this book has numerous tools to support teacher learning: classroom vignettes, lessons, and assessments, tools for self-reflection, tools for professional development, ways to work with communities, and discussion questions.

Dacey, L., Lynch, J. B., & Salemi, R. E. (2013). *How to Differentiate Your Math Instruction: Lessons, Ideas, and Videos With Common Core Support, Grades K–5.* Sausalito, CA: MathSolutions.

This book provides a wealth of information and strategies for implementing differentiated instruction. In addition to attention on adapting tasks, ideas for engaging students, providing choice, scaffolding learning, and managing the classroom are provided. Each of these topics includes vignettes, videos, reflection guides, and downloadable tools.

Small, M. (2017). *Good Questions: Great Ways to Differentiate Mathematics Instruction in the Standards-Based Classroom* (3rd ed.). Reston, VA: NCTM.

Small, M., & Lin, A. (2010). *More Good Questions: Great Ways to Differentiate Secondary Mathematics Instruction.* Reston, VA: NCTM.

These books provide a wealth of great examples of differentiated tasks, organized by content area and then by open tasks and parallel tasks. There are many great ideas, including purposeful questions to accompany the tasks. And the third edition of Good Questions *(2017) maps the tasks to CCSS-M content and Mathematical Practices.*

Smith, N. (2017). *Every Math Learner: A Doable Approach to Teaching and Learning With Differences in Mind, Grades K–5.* Thousand Oaks, CA: Corwin.

Smith, N. (2017). *Every Math Learner: A Doable Approach to Teaching and Learning With Differences in Mind, Grades 6–12.* Thousand Oaks, CA: Corwin.

These two books provide a comprehensive approach to differentiating instruction. The authors include strategies for determining learning profiles for students, as well as ideas for planning, implementing, and assessing differentiated tasks.

Articles

Drake, C., Land, T. J., Bartell, T. G., Aguirre, J. M., Foote, M. Q., McDuffie, A. R., & Turner, E. E. (2015). "Three Strategies for Opening Curriculum Spaces." *Teaching Children Mathematics, 21*(6), 346–353.

This article describes three ways to connect content to the learner: meaning-making in a lesson by rearranging lesson components, adapting tasks, and making authentic connections. They are all ways to differentiate instruction and better meet the needs of all learners.

Hand, V., Kirtley, K., & Matassa, M. (2015). "Narrowing Participation Gaps." *Mathematics Teacher, 109*(4), 262–268.

These authors challenge us to consider the kinds of opportunities we create for different groups of students to participate in reasoning and argumentation activities. They provide a compelling discussion of complex instruction, including three important overarching ways teachers strive to engage every student.

Holden, B. (2008). "Preparing for Problem Solving." *Teaching Children Mathematics, 14*(5), 290–295.

A first-grade teacher in an urban, high-poverty setting incorporated differentiated instruction with great results. Holden describes six specific strategies she used to help her students to be successful.

Kabiri, M. S., & Smith, N. L. (2003). "Turning Traditional Textbook Problems Into Open-Ended Problems." *Mathematics Teaching in the Middle School, 9*(3), 186–192.

The title says it all—four examples are provided (one in each content area). This article (appropriate for K–12 teachers) provides strong support for helping teachers see how they can get more out of the resources they are using.

Williams, L. (2008). "Tiering and Scaffolding: Two Strategies for Providing Access to Important Mathematics." *Teaching Children Mathematics, 14*(6), 324–330.

This article describes how two lessons (second-grade fractions and third-grade geometry) were tiered and then how scaffolds were built into the lesson. It is a great article to discuss with colleagues as a launch into how to differentiate other mathematics lessons.

Online Resources

IllustrativeMathematics

https://www.illustrativemathematics.org/content-standards

While these lessons do not provide explicit strategies for differentiation, there are many excellent examples of tasks that are open and that have features of differentiated tasks (e.g., students selecting their own representation, strategy, or pathway).

Steve Leinwand

http://steveleinwand.com

Among other useful resources, Steve maintains a downloadable list of "Great Resources." On the list are numerous websites that offer quality lessons and tasks that can be used to discuss how they reflect a differentiated approach, as well as how they might be adapted to be parallel or tiered tasks.

Coach's Toolkit

These tools are a menu from which you can select any that make sense for your setting/context. They can be used independently or as part of a coaching cycle. You may start with the self-assessment, which can guide you in deciding which of the other tools may be most useful. If using these tools for a coaching cycle, mix and match as you like or use one of the combinations we suggest in the diagrams that follow. The tools in this chapter include instructions to the coach and the teacher. You can download copies of the tools that only have instructions for the teacher at **resources.corwin .com/mathematicscoaching.**

Self Assess

8.1	Connecting *Shifts* to Differentiated Instruction Self-Assessment

Gather Data

8.6	Focus on Five
8.7	Differentiating Instruction

Plan

8.2	Student Diversity
8.3	Meeting Individual Needs
8.4	Different Ways to Differentiate a Lesson
8.5	Opening Closed Tasks

Reflect

8.8	Learning From Focus on Five
8.9	Impact of Open/Tiered/ Parallel Tasks

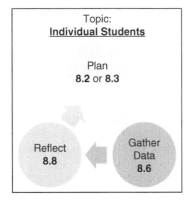

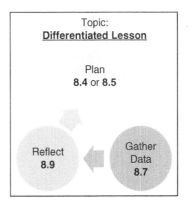

Additional Tools in Other Chapters

Tools in Chapters 9 and 10 focus specifically on emergent multilingual learners and students with special needs. Tool 3.6, Worthwhile Task Analysis, can also be used to focus on differentiating tasks. Tools in Chapter 6 (Formative Assessment) and Chapter 7 (Analyzing Student Work) can also be used to focus on the impact of tasks on each student's learning.

 To download the coaching tools for Chapter 8 that only have instructions for the teacher, go to **resources.corwin.com/mathematicscoaching.**

8.1 Connecting *Shifts* to Differentiated Instruction Self-Assessment

Instructions to the Coach: Ask teachers (individually or as part of a PLC activity) to self-assess where they position themselves on each of these Shifts in Classroom Practice *related to differentiated instruction. Use the following questions during a coaching conversation or in a PD setting to support teachers (and to help you decide which tools may be most useful from this chapter). A one-page version of this tool without this note is available for download.*

Instructions: The *Shifts in Classroom Practice* listed below have specific connections to differentiating instruction. Put an *X* on the continuum of each *Shift* to identify where you currently see your practice.

Tool 8.1 Shifts

Shift 1: From *stating-a-standard* toward *communicating expectations for learning*

Teacher shares broad performance goals and/or those provided in standards or curriculum documents. ⟶ Teacher creates lesson-specific learning goals and communicates these goals at critical times within the lesson to ensure students understand the lesson's purpose and what is expected of them.

Shift 3: From *teaching about representations* toward *teaching through representations*

Teacher shows students how to create a representation (e.g., a graph or picture). ⟶ Teacher uses lesson goals to determine whether to highlight particular representations or to have students select a representation; in both cases, teacher provides opportunities for students to compare different representations and how they connect to key mathematical concepts.

Shift 7: From *mathematics-made-easy* toward *mathematics-takes-time*

Teacher presents mathematics in small chunks so that students reach solutions quickly. ⟶ Teacher questions, encourages, provides time, and explicitly states the value of grappling with mathematical tasks, making multiple attempts, and learning from mistakes.

Shift 8: From *looking at correct answers* toward *looking for students' thinking*

Teacher attends to whether an answer or procedure is (or is not) correct. ⟶ Teacher identifies specific strategies or representations that are important to notice; strategically uses observations, student responses to questions, and written work to determine what students understand; and uses these data to inform in-the-moment discourse and future lessons.

Tool 8.1 Reflection Questions

1. What do you notice, in general, about your self-assessment of these *Shifts in Classroom Practice*?

2. What might be specific teaching moves that align with where you placed yourself on the *Shifts*?

3. What might be specific teaching moves that align *to the right of* where you placed yourself on the *Shifts*?

4. What might be some professional learning opportunities to help you move to the right for one or more of these *Shifts*?

8.2 Student Diversity

Instructions to the Coach: This tool can be paired with 8.4 or 8.5 to help a teacher consider possible strategies. Focus attention on whether the strategies maintain high expectations.

Instructions: As you prepare for a lesson, think about each student individually to be sure all have access to the strategies you are planning to use.

Subgroups	Number of Students	Possible Strategies
	Gather this data prior to a planning conversation.	List adaptations or instructional strategies you can use to make mathematics accessible and meaningful to each subgroup.
Girls/Boys		
Students with learning disabilities		
Students from various culture/ethnic groups		
Emergent multilinguals		
Gifted or advanced students		
Struggling or reluctant students		

Planning Questions

1. In general, what strategies might you use to meet the needs of the diverse learners in your classroom? (How might you plan a lesson to meet the needs of every child?)

2. For this lesson/content specifically, what different/additional strategies will you use related to supporting the learners identified in the table? How might grouping be used to support all learners?

3. How might you monitor whether each student is understanding? Participating?

4. What strategies might you use to address students who are struggling? Giving up? Finding the task too easy?

Previously published by Bay-Williams, J., McGatha, M., Kobett, B., and Wray, J. (2014). Mathematics Coaching: Resources and Tools for Coaches and Leaders, K–12. New York, NY: Pearson Education, Inc.

8.3 Meeting Individual Needs

Instructions: Differentiation begins with meeting individual needs. Consider the strengths of each individual (resources) and learning preferences. Then, consider ways to adapt the lesson while maintaining the lesson goals and rigor.

Name of Student	Resources and Needs	Differentiation Strategy

Planning Questions

1. What strengths does each student bring to class in terms of content knowledge and in terms of his or her home/community experiences?

2. How might these strengths/interests be integrated into the lesson to increase access, motivation, and learning?

3. In what ways might the differentiation strategies ensure that each person is a critical member of the productive classroom culture?

8.4 Different Ways to Differentiate a Lesson

Instructions to the Coach: This tool can be used in professional learning, PLCs, or a coaching cycle.

Instructions: Use one of these three approaches to differentiating a lesson:

- Content (what you want each student to know)
- Process (how you engage students in thinking about that content)
- Product (how students will demonstrate what they have learned)

Record possible ways to differentiate a selected task or lesson. Select one or more to implement.

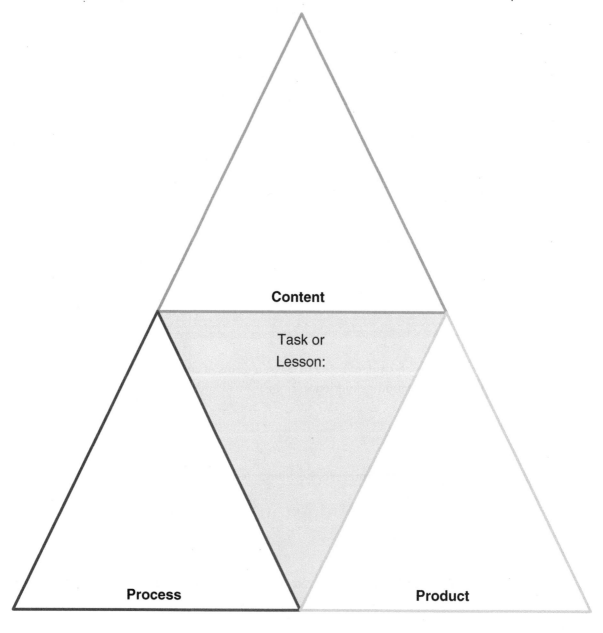

Content

Task or
Lesson:

Process   Product

8.5 Opening Closed Tasks

Instructions to the Coach: This tool can be used in professional learning, PLCs, or a coaching cycle.

Instructions: Complete either or both of the adapting tasks activities.

Opening a Task. Select a closed task (e.g., from a textbook or any other resource). Small and Lin (2010) and Small (2017) suggest the following strategies for opening a task. Use one of these or any other.

Turning around a question	Asking for similarities and differences
Replacing given information (e.g., number, unit, shape) with a blank	Writing a sentence with selected words/numbers in it

My opening task strategy: _____

Before	After

Tiered or Parallel Tasks. Select a task (open or closed) and adapt it into two or three parallel tasks that meet the lesson objectives but vary in other ways. Choose one of these ways (or any other): difficulty, structure, or design.

My parallel task strategy: _____

Before	After

8.6 Focus on Five

Instructions to the Coach: Prior to the data gathering, ask the teacher for the names of up to five students to observe. Record the students' engagement (middle column) and the teacher moves (right column) related to that student.

Name of Student	Student Participation and Actions	Teacher Actions

 Available for download at **resources.corwin.com/mathematicscoaching.** Copyright © 2018 by Corwin.

8.7 Differentiating Instruction

Instructions to the Coach: Gather data based on which type(s) of differentiation a teacher has selected (e.g., content). Record any related statements, questions, classroom organization, and/or management.

Type of Differentiation		Teacher
Content	**What you want each student to know**—includes adapting the types of problems students are doing, changing objectives, working on different content in different groups, and so on. Are all students exploring content that is appropriate for their age and prior knowledge?	
Process	**How you engage students**—includes grouping strategies, what is done in introducing a task to make it comprehensible, tools that are used, lesson design, and so on. Are different strategies used to ensure all students understand and can participate in the task?	
Product	**How students will demonstrate what they have learned**—includes what students do during the lesson (record in a journal, explain to a peer, create a graph) and upon completion of the lesson. Are choices offered or different ways encouraged?	

Previously published by Bay-Williams, J., McGatha, M., Kobett, B., and Wray, J. (2014). Mathematics Coaching: Resources and Tools for Coaches and Leaders, K–12. *New York, NY: Pearson Education, Inc.*

8.8 Learning From Focus on Five

Instructions to the Coach: This tool can be paired with 8.6 to help a teacher focus on specific data from individual students.

Instructions: Describe insights from the data on selected students as a way to gain insights for supporting every student.

1. Before the lesson, what did you perceive to be the specific learning needs of this student related to the learning objectives?

2. In looking at the data that was gathered, what successes and what struggles did this student encounter in the lesson?

3. What structures or strategies were in place that seemed to best support each child's learning? What strategies were not effective?

4. To what extent did these students do the following:

   Learn the objectives

   Persevere in solving the problems

   Participate in and contribute to the mathematical community

5. What next steps might you take to address any or all of the items listed in #4?

6. What are you learning about making the content accessible, challenging, and meaningful for the selected students? For all students?

8.9 Impact of Open/Tiered/Parallel Tasks

Instructions to the Coach: This tool can be used in a PLC, with you providing student work or teachers bringing their own students' work. Or it can be used to reflect on a lesson that used an open, tiered, or parallel task.

Instructions: After analyzing data from implementing an open, parallel, or tiered task, sort student work in terms of the extent that they demonstrated understanding of the lesson goals.

About Students . . .

1. What do you notice about the students who successfully demonstrated understanding?

2. In what ways did opening the task or creating tiered/parallel options impact student engagement and perseverance?

3. What were the benefits (and issues) for the range of learners in your classroom (i.e., in what ways did the task(s) benefit different groups of students)?

About Teaching . . .

4. What challenges did you encounter in the implementation of the (open/tiered/parallel) task?

5. In what ways did the use of (open/tiered/parallel) tasks allow you to observe and support student thinking?

6. What about adapting tasks do you want to remember in terms of planning such tasks? Implementing such tasks?

Chapter 9

Supporting Emergent Multilingual Students

Dear Coach,

Here is support for you as you work with teachers on supporting emergent multilingual students.

In the COACH'S DIGEST ...

Overview: Strategies for supporting both mathematical and language development for emergent multilingual students to read and/ or to download and share with teachers.

Coaching Considerations for Professional Learning: Ideas for how to support teachers in considering strategies to support emergent multilingual students.

Coaching Lessons From the Field: Mathematics coaches and leaders share how they have supported emergent multilingual students, using ideas and tools in this chapter.

Connecting to the Framework: Specific ways to connect *Shifts* and Mathematical Practices to emergent multilingual students.

Coaching Questions for Discussion: Menu of prompts for professional learning or one-on-one coaching related to teaching emergent multilingual students.

Where to Learn More: Articles, books, and online resources for you and your teachers to learn more on this topic!

In the COACH'S TOOLKIT ...

Eight tools focused on equitable mathematics teaching, for professional learning or coaching cycles.

Coach's Digest

In the Coach's Digest, we begin with an overview of supporting emergent multilingual students, written to teachers (and to you). As mentioned in Chapter 8, we use the phrase *emergent multilingual student* instead of the more commonly used English language learner (ELL) or English learner (EL) because it is more accurate terminology. As you read the Overview, the following questions might help you reflect on this topic in terms of your role as a mathematics coach:

• If I build on my teachers' strengths, as I am asking them to do with their students, where might I start in implementing these strategies?

• What types of professional learning activities, readings, and/or coaching cycles/lesson study might support their efforts to implement these ideas?

• What data might we collect to see if the strategies are indeed supporting emergent multilingual students?

Overview of Supporting Emergent Multilingual Students

online resources ↗ To download the Chapter 9 Overview, go to **resources.corwin.com/mathematicscoaching.**

Students' native languages are not only an important part of their cultural heritage, but they are vital to how they think, communicate, and learn. We must draw on students' funds of knowledge, which includes building on their cultural and community knowledge, recognizing their language and cultural heritage as a resource, and connecting mathematics instruction to their experiences and interests (Aguirre, Mayfield-Ingram, & Martin, 2013; Bartell et al., 2017). Comparing algorithms from different countries, exploring games or cooking from different regions, and storytelling can serve as cultural and linguistic resources in learning mathematics.

A quick review of the Mathematical Practices illuminates the critical role that language plays in learning mathematics. Student actions described in these eight Practices include making conjectures, reasoning, explaining connections among representations, constructing arguments, justifying conclusions, listening to the arguments of others, asking questions, and communicating precisely to others. Listening, speaking, reading, and writing—all embedded in the Mathematical Practices to focus on the learning of mathematics—are also the language processes needed to support the development of language. **Culturally responsive mathematics instruction (CRMI)** focuses on the big ideas of mathematics, makes that content relevant to the student, attends to each student's identity, and ensures each student's contributions are "taken up" in the classroom in an equitable manner. With this as the goal for supporting all students, here we focus specifically on effective strategies for supporting emergent multilingual students (as we stated, the phrase *emergent multilingual student* replaces *English language learner* because it is a more accurate and inclusive phrase). The following 10 strategies are not meant to be comprehensive, but to highlight critical strategies, based on research, that support emergent multilingual students.

1. *Communicate high expectations*. How do you respond when an emergent multilingual student does not start to work on a problem or assignment? Too often, our first attempt to "help" is to simplify the mathematics and/or remove the language from the lesson. Both of these teacher moves can instead eliminate opportunities to develop mathematical proficiency. Instead, focus on the big ideas of mathematics and use tasks worthy of group work. Such tasks emphasize multiple representations and incorporate students' justifications and presentations, which support equitable opportunities to learn math (Nasir, Cabana, Shreve, Woodbury, & Louie 2014; Dunleavy, 2015).

2. *Make content relevant*. This includes two equally important elements. First, is the *mathematics itself* presented meaningfully, and is it connected to other content? Second, are the situations or contexts familiar and interesting?

3. *Establish norms for participation*. All students deserve opportunities to develop a strong mathematics identity (Bartell et al., 2017). A mathematical identity emerges from how a students' mathematics ideas and contributions are received by their peers and their teachers. Therefore, effective teaching of emergent multilinguals includes explicit attention to establishing norms that value all students' contributions and equitable participation among classmates, as well as establishes positive social interactions in small groups.

4. *Honor native language and other languages*. Students can communicate in their native language as they learn mathematics and still advance their English proficiency (Haas & Gort, 2009; Moschkovich, 2009). Students might do their initial thinking in small groups in their preferred language and then consider how to communicate those ideas to

others by translating their ideas into English or by using visuals or pictures. Multilingual students often code-switch, or move between two languages. Code-switching supports mathematical reasoning because students are selecting the language in which they can best express their ideas (Moschkovich, 2009). Explicitly highlighting connections between languages is also interesting and helpful to students. Many mathematics words are similar across languages (Celedón-Pattichis, 2009; Gómez, 2010)—for example, *equal* (English) and *igual* (Spanish). If you know these words, share them. If you don't, ask your students (or use a translator app).

5. *Set content and language goals.* The first Teaching Practice (NCTM, 2014) states that goals should "identify the Mathematical Practices that students are learning to use more proficiently." These include significant use of language. An objective such as "Students will describe in writing how two strategies compare to each other" attends to both mathematical and language outcomes for students. And having such an objective means teaching must attend not only to using multiple strategies but also to what counts as a difference or similarity and how that might be written or stated.

6. *Provide vocabulary support.* Language support benefits every student, but it is absolutely essential in accomplishing #3 in this list. Can you think of a word that means something different in mathematics classrooms than it typically means outside of mathematics classroom? There are many (e.g., *sum (some), mean, table, difference, etc.*). In fact, it is likely that any given lesson has at least one word that is taking on specialized meaning. Auditing lessons for both context and content vocabulary is important. If a context is not familiar to students, change it or preteach the necessary context words. While a context is critical to building understanding, too many contexts can take away from learning. Consider staying with one context for situations rather than using a variety of contexts. For example, for multiplication story problems, use stories about the same situation but vary the missing values in the story. This removes unnecessary language demands while maintaining high-level mathematical thinking.

7. *Use comprehensible input.* This is closely related to vocabulary support, but it is a more general focus on communicating ideas so that they are understood. For example, giving instructions using visual cues can help an emergent multilingual student understand the expectations. Visuals, realia, picture books, and many other resources can be used to ensure the student understands the big ideas of the lesson and the expectations for doing a task.

8. *Use cooperative groups strategically.* Building on #3 in this list, it is important to use cooperative groups with emergent multilinguals (Baker et al., 2014). Grouping of students must consider students' language skills. Placing a student who is just beginning to learn English with two English-speaking students may result in the student being left out, but grouping all Spanish speakers together prevents the students from learning English. Try to place bilingual students with students who are just beginning to learn English. Emphasize that using different languages in small groups is acceptable, but they must make sure everyone in their group understands and is engaged in the conversations.

9. *Select tasks with multiple entry and exit points.* When a task is open-ended, students can apply their prior knowledge and experiences to solve the task. Algorithms vary from country to country, as do ways to represent mathematics concepts. Inviting and encouraging different methods and representations will result in approaches to problems that may be novel to other students (and the teacher!). This also presents an exciting opportunity for engaging the class in seeing how different strategies are alike and different.

10. *Use diagnostic assessment tools*. **Diagnostic interviews** provide a chance to observe what content or language a student does or does not understand. Too often, assumptions are made about students' *mathematical understanding* when in fact the challenge was *linguistic*. Watching students solve a problem and probing to see what they know about the problem, as posed, can provide important insights into their mathematical and language knowledge, which can then inform instructional decisions.

Across these 10 strategies is a consistent and critical message: Every student must have the opportunity to develop a positive mathematics identity, and cultural diversity in a classroom can enrich the learning of mathematics.

NOTES

_____

_____

_____

_____

_____

_____

_____

_____

_____

_____

_____

_____

_____

_____

_____

_____

_____

_____

_____

_____

_____

_____

_____

_____

_____

_____

_____

_____

_____

Coaching Considerations for Professional Learning

Teachers vary in their experience with emergent multilingual students, with other languages, and with the strategies they might use. A very experienced teacher may have two emergent multilingual students for the first time in her career, or a first-year teacher may have a classroom full of emergent multilinguals. In any case, finding out what experiences they have had models what we hope they will do with their students. Here are some ideas for working with teachers to increase their capacity to support and challenge emergent multilingual students in learning important mathematics.

1. *Analyze tasks with a dual focus on mathematics and language.* Teachers are accustomed to looking at a problem and thinking of the prior content knowledge needed to engage in the problem, but they are less familiar with considering the language needs. Language "noticings" might include phrasing, such as if–then statements, or terms associated with the context of the problem. Teachers may gloss over a lesson that uses four terms for the same thing (e.g., school lot, school grounds, yard, or property), not realizing the unnecessary linguistic challenge this causes. A protocol might first focus on what *mathematics* might be needed to solve the problem. At times, teachers will offer suggestions that are not really needed to solve the problem (knowing an algorithm, for example, because the task could be solved through reasoning). Second, consider language: Is the context familiar? Should it be adapted? Could rewriting the sentences help with comprehension? Are there terms or phrases that could be misunderstood? Third, decide what attention to mathematical and contextual language belongs at the start of the lesson and which terms make more sense to save for the discussion later in the lesson. The list must be short! Or review a lesson they are teaching and use Tool 9.3.

2. *Co-construct lessons.* Chapter 8 describes three ways to differentiate a task (content, process, and product; see also Planning Tool 8.4). Consider using these ideas of differentiation targeting the specific needs of emergent multilingual students. For elaboration on these specific strategies, use the Overview or select from the list of recommended readings. Key to this discussion and the effort to modify instruction is to be sure that the modified tasks still require a high level of cognitive demand. Removing the story situations, a popular strategy for reducing the difficulties for emergent multilinguals, may in fact remove (a) the challenge or (b) the connection to a meaningful context that can help them understand the abstract symbols.

3. *Explore a student's language comprehension.* Some emergent multilingual students have a very strong understanding of everyday language, so the teacher may assume that the academic language (about the mathematics) is understood. But the unfamiliar sentence structuring and vocabulary in mathematics can continue to limit comprehension. Discuss with teachers how they might know whether an emergent multilingual student understands a term, a concept, or a task he or she is to solve. Identify a tool from Chapter 6 (Formative Assessment) to use.

4. *Co-develop diagnostic interviews.* Research indicates that when an emergent multilingual student's assignment is incorrect or incomplete, teachers tend to assume lack of content knowledge when the issue is often language related. Interviewing emergent multilingual students can address this issue. Co-designing a diagnostic interview for a student or students can be part of a coaching cycle. Data gathering can be done by the teacher (via the interview), or the coach can observe the interview and take notes on the strategies used by the teacher. Reflection discussions can focus on (1) what the student did or didn't know and (2) how effective the interview was or wasn't (see Tools 9.4 and 9.8).

5. *Help teachers ensure participation.* Participation includes consideration of language and culture. Building a community in which an emergent multilingual student is given opportunities to practice language use in nonthreatening situations and in public ways is critical. With teachers, you can discuss some of the factors that they must take into consideration. For example, students from some countries may not feel comfortable critiquing the reasoning of others (Mathematical Practice 3), so teachers need to model what this looks like and help students practice. Talk moves, such as revoicing, become even more critical as a tool to make sure students follow the conversation. Discuss what structures might help a small group value each student's contributions (see Tools 9.5 and 9.6).

6. *Focus on culturally responsive mathematics instruction.* Contrary to what some may think, mathematics is not universal—it is a cultural practice. There are numerous ways to develop instruction that is culturally responsive. The extensive literature on this topic provides many ideas that tend to fall under four categories: (1) important content, (2) relevant content, (3) positive mathematical identities, and (4) valuing each student's contributions. Tool 9.2 provides a series of reflective questions focused on CRMI, and this tool, along with Tools 9.5 and 9.7, can be used for a coaching cycle.

Coaching Lessons From the Field

I like the worthwhile task analysis tool (Tool 3.6), but I wanted to shorten it and add a focus on social justice because we were working on trying to connect mathematics to students' lives. I adapted the task analysis tool with the intent to help teachers either find or develop appropriate tasks for the grade level and maturity of students without compromising the mathematics (see Figure 9.1). I have used it in professional learning to discuss potential tasks to address socially relevant topics and then to consider ways we can enhance or change a task so that it does.

–Linda Fulmore, Education Consultant
Cave Creek, AZ

| Figure 9.1 | Adapted Tool to Include a Focus on CRMI, Including Social Justice |

Indicator	Little or No Focus (a rating of 1)	Rating			The Expectation (a Rating of 3)
		1	2	3	
1. *The task is grade-level appropriate.* The task must be identifiable in the curriculum, with the right amount of challenge.	The task is below the grade-level curriculum.				The task represents an expectation of the grade-level curriculum.
2. *The task makes connections between concepts and procedures.* The task must be in the category of *Procedures With Connections* or *Doing Mathematics* as identified by Stein and Smith (2011).	The task focuses on procedures. The focus is on a formula or series of steps to a correct answer.				The task requires conceptual thinking and understanding to broaden mathematics knowledge, leading to deeper understanding about the event.

Figure 9.1	Adapted Tool to Include a Focus on CRMI, Including Social Justice *(Continued)*					

Indicator	Little or No Focus (a rating of 1)	Rating			The Expectation (a Rating of 3)
		1	2	3	
3. *The task is designed to engage the students' interest and intellect on a social injustice event.* Secondary students can conceptualize poverty and incarceration rates by population—unless these events represent their lived experiences (then, teachers must proceed with caution).	The task is based on a social justice event inappropriate for the maturity of students or lived experiences.				Students are mature enough to show interest, curiosity, and empathy as discussions unfold.
4. *The task provides varying entry points.*	The task does not require complex thinking. The goal is to arrive at a solution from teacher-given information.				The task is presented as a statement or question requiring inquiry. Students decide what data is needed to answer the question.
5. *The task can offer students opportunities to reflect, illustrate, explain, or justify thinking about a social justice event.* These opportunities are communicated in the lesson plan.	The task requires a mathematics solution only.				The task requires students to use mathematics learning to make sense of and form opinions about an event. Students can determine trends.

Connecting to the Leading for Mathematical Proficiency (LMP) Framework

As teachers focus on supporting emergent multilingual students, it is important for them to make explicit connections to the *Shifts in Classroom Practice* and the Mathematical Practices. The brief paragraphs that follow provide ideas for making these connections. You and your teachers can continue to add to these ideas! Tool 9.1 can also be used for connecting to the *Shifts*.

Connecting Supporting Emergent Multilingual Students to *Shifts in Classroom Practice*

Shifts   1   2   3   4   5   6   7   8

Emergent multilingual students bring both linguistic and cultural resources to the classroom. These resources are accessed and their language supported when classrooms reflect the right sides of the *Shifts*.

Shift 1: From *stating-a-standard* toward *communicating expectations for learning*

Teacher shares broad performance goals and/or those provided in standards or curriculum documents. ⟶ Teacher creates lesson-specific learning goals and communicates these goals at critical times within the lesson to ensure students understand the lesson's purpose and what is expected of them.

Non-native English speakers miss many parts of the conversation. When teachers are very intentional about making the big ideas of the lesson clear to students, these missed details don't negatively impact the student. Revisiting these big ideas throughout the lesson helps emergent multilingual students follow and participate in the lesson.

Shift 2: From *routine tasks* toward *reasoning tasks*

Teacher uses tasks involving recall of previously learned facts, rules, or definitions and provides students with specific strategies to follow. ⟶ Teacher uses tasks that lend themselves to multiple representations, strategies, or pathways encouraging student explanation (how) and justification (why/when) of solution strategies.

As described in the Overview, it is not appropriate to move to the left on this continuum for an emergent multilingual student. All students benefit from teaching that lies in the right end, complete with tasks that include contexts and complexities. Emergent multilinguals, in particular, benefit from using multiple representations and pathways, which invite them to use strategies from their native country and/or their personal experiences.

Shift 3: From *teaching about representations* toward *teaching through representations*

Teacher shows students how to create a representation (e.g., a graph or picture). ⟶ Teacher uses lesson goals to determine whether to highlight particular representations or to have students select a representation; in both cases, teacher provides opportunities for students to compare different representations and how they connect to key mathematical concepts.

As stated in *Shift* 2, emergent multilingual students benefit from representations. These visuals not only enable them to make the mathematical connections but to understand and use the specialized language of mathematics. As students are invited to compare representations, emergent multilinguals are given an opportunity to use language and to showcase thinking that may be quite different from their peers'.

Shift 4: From *show-and-tell* toward *share-and-compare*

Teacher has students share their answers. ⟶ Teacher creates a dynamic forum where students share, listen, honor, and critique each other's ideas to clarify and deepen mathematical understandings and language; teacher strategically invites participation in ways that facilitate mathematical connections.

As described in the 10 strategies to support emergent multilingual students, norms must be established to value everyone's contributions and to ensure that all students develop a positive mathematics identity. The teacher must invite participation and position each student as capable.

Shift 6: From *teaching so that students replicate procedures* toward *teaching so that students select efficient strategies*

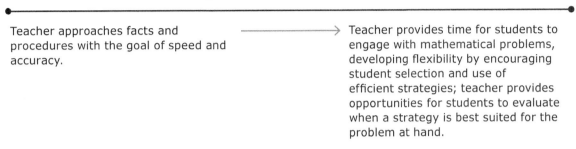

Teacher approaches facts and procedures with the goal of speed and accuracy.	→	Teacher provides time for students to engage with mathematical problems, developing flexibility by encouraging student selection and use of efficient strategies; teacher provides opportunities for students to evaluate when a strategy is best suited for the problem at hand.

Standard algorithms are not the international standard. Many algorithms from other countries are very efficient and less prone to error than many US algorithms or strategies. When a teacher encourages students to select efficient strategies and solicits algorithms or strategies from students' home cultures, every student in the class benefits.

> While these four *Shifts* were identified as particularly important, every *Shift* must take into consideration how to make it work for every student. When those students who are learning English are also learning mathematics, additional considerations are needed. Help teachers see how to implement these additional strategies while still moving to the right on the *Shift* by making this point and providing a forum for teachers to figure out how to do this well.

Connecting Supporting Emergent Multilingual Students to Mathematical Practices

MPs   **1**   2   3   4   **5**   **6**   7   8

All students must have opportunities to become mathematically proficient, meaning that they can exhibit all of the Mathematical Practices. We have, however, identified three particularly powerful Mathematical Practices that support emergent multilingual students in learning mathematics.

1. **Make sense of problems and persevere in solving them**. If you sort of know a second language but are not fluent in that language, try to count to 20 and explain how you solved $6\frac{1}{8} - 4\frac{3}{4}$. If the counting task is easy for you but you can't explain your strategy for subtracting the fractions, you can begin to understand the difficulty that students learning English face in a mathematics classroom. Unfortunately, when they cannot explain—or wish not to try because it is so hard to find the words to do it—assumptions can be made about their mathematical understanding that underestimate their ability and understanding.

5. **Use appropriate tools strategically**. Tools (calculators, manipulatives, digital content, etc.) provide visuals that help all students focus on important mathematical ideas. For students who may not have the academic vocabulary, tools also assist in understanding a worthwhile task that has been presented to them. In addition, tools provide a way in which they can demonstrate their knowledge when they do not have the language they need to share their strategy. Using digital content is also a way to support the learning of emergent multilingual students, as they can find websites in their native language that can be applied to the problem they are solving. Emergent multilinguals may choose a different tool, one that is more familiar to them based on their culture or experiences. Encouraging students to consider and select appropriate tools honors their backgrounds and develops this mathematical proficiency.

6. **Attend to precision**. All students benefit from explicit attention to vocabulary. When the teacher attends to appropriate terminology, students use the appropriate terminology. For emergent multilingual students, increased attention is needed for using vocabulary. The extent of language support depends on whether or not the student has an understanding of the term in his or her native language. Visuals and translations can aid in developing vocabulary. In addition, measurement units in other countries are metric units. When addressing precision of units, students from other countries will likely be stronger than US natives in considering the size of units when using metric measures, yet may struggle more with US standard measures.

> The Mathematical Practices are actually a "good news" story. The very things they recommend align well with equitable instruction. Teachers, however, may be concerned about the linguistic demands embedded within the Mathematical Practices. Sharing connections to research on effective practices for emergent multilingual students can help. Additionally, teachers may have students who are have not been formally schooled and therefore are not ready for the content that is in the curriculum. Help them to see how they might engage these students in the Mathematical Practices, but with content at their readiness level.

NOTES

_____

_____

_____

_____

_____

_____

_____

_____

_____

_____

_____

_____

Coaching Questions for Discussion

Questions Related to the Focus Zone: Supporting Emergent Multilingual Students

In addition to the questions in Chapter 8, these questions can engage teachers in thinking about supporting emergent multilingual students.

1. What resources does each student bring to learning mathematics (experiences, cultural heritage, language)? How might you learn more about these? How might these resources be accessed in the lesson?

2. What might be some ways to adapt a task/lesson so that it is more responsive to your emergent multilingual students?

3. As you think about adapting tasks, how can you make them more accessible and yet maintain a high level of thinking?

4. What actions might be taken in the introduction of a lesson to ensure everyone understands the big idea of the lesson and what the problem is asking students to do? (How might you build background?)

5. In structuring small-group discussions, how might you ensure that every student is a full participant?

6. In whole-class discussions, how might you ensure that every student is a full participant?

7. As a lesson is "in progress," what strategies might you use to gauge whether a student comprehends what is happening?

8. How might you determine whether your emergent multilingual students understand what has been presented? If the work is incomplete, how might you sort out whether the issue is language related or content related?

Questions Related to the LMP Framework

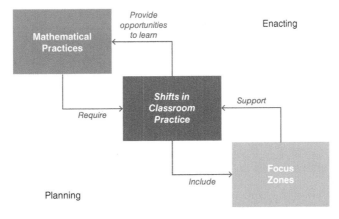

1. As you enact the ideas of supporting emergent multilingual students, what do you see as related *Shifts in Classroom Practice*?

2. As you enact the ideas of supporting emergent multilingual students, what do you see as opportunities to learn and, in particular, to develop the Mathematical Practices?

Where to Learn More

Books

Aguirre, J. M., Mayfield-Ingram, K., & Martin, D. (2013). *The Impact of Identity in K–8 Mathematics Learning and Teaching: Rethinking Equity-Based Practices.* Reston, VA: NCTM.

This book describes five practices for equitable mathematics teaching, with differentiation themes embedded. Numerous tools to support teacher learning can be a great support to mathematics coaching: classroom vignettes, lessons and assessments, tools for self-reflection, tools for professional development, ways to work with communities, and discussion questions.

Civil, M., & Turner, E. (2014). *Common Core State Standards in Mathematics for English Language Learners: Grades K–8.* Alexandria, VA: TESOL Press.

With a rich collection of classroom vignettes, the authors illustrate teaching practices that support emergent multilingual students in learning the Mathematical Practices.

Kersaint, G., Thompson, D. R., & Petkova, M. (2009). *Teaching Mathematics to English Language Learners.* New York: Routledge.

This is an excellent resource for learning about the special mathematics-related considerations for emergent multilingual students. So many specific examples are provided across the grades, it gives much support for all grade levels.

Ramirez, N., & Celedón-Pattichis, S. (2012). *Beyond Good Teaching: Advancing Mathematics Education for ELLs.* Reston, VA: NCTM.

This is a wonderful resource of research-informed, pragmatic strategies to support emergent multilingual students. It presents principles for teaching, along with specific student and teacher actions for different levels of language development. You will find messages from students and teachers, as well as cases of practice with reflective questions. Online resources include videos, observation and analysis protocols, and lesson planning tools.

Articles

Banse, H. W., Palacios, Merritt, E. G., & Rimm-Kaufman, S. E. (2016). "Five Strategies for Scaffolding Mathematical Discourse With ELLs." *Teaching Children Mathematics, 23*(2), 100–108.

This article provides five effective strategies for ensuring that students who are learning English are full participants in classroom discussions. Teacher moves and vignettes illustrate each strategy, making it a great article to share with teachers or to read for your own professional learning.

Bay-Williams, J. M., & Livers, S. (2009). "Math Vocabulary Acquisition." *Teaching Children Mathematics, 16*(4), 238–246.

Livers, S. D., & Bay-Williams, J. M. (2014). "Timing Vocabulary Support: Constructing (Not Obstructing) Meaning." *Mathematics Teaching in the Middle School, 10*(3), 153–159.

These articles focus on how (and when) to provide vocabulary support in a lesson in ways to keep the mathematical challenge high. Many activities for developing vocabulary are offered.

Thompson, D., & Rubenstein, R. (2000). "Learning Mathematics Vocabulary: Potential Pitfalls and Instructional Strategies." *Mathematics Teacher, 93*(7), 568–574.

In an at-a-glance table, the authors share different kinds of language pitfalls. This is followed by oral, written, and visual strategies for supporting vocabulary development. A good read for all teachers.

Whiteford, T. (2009/2010). "Is Mathematics a Universal Language?" *Teaching Children Mathematics, 16*(5), 276–283.

This article reviews the many ways that procedural knowledge varies across languages and cultures. A great read for elementary teachers.

Online Resources

Scaffolding Instruction for English Language Learners

https://www.engageny.org/resource/scaffolding-instruction-english-language-learners-resource-guides-english-language-arts-and

This link takes you to a landing page for EngageNY, but on that page is a document (or PDF) titled Scaffolding Instruction for ELLs: Resource Guide for Mathematics. *This practical guide, developed with the support of the American Institute for Research, provides strategies that are relevant to any mathematics lesson (and connected to the CCSS-M) to support emergent multilinguals. The samples provide good ideas for how to apply the language supports to other mathematics lessons.*

Teaching Channel

https://www.teachingchannel.org/blog/categories/english-language-learners

The blogs posted here (written by teachers) can be great resources for launching discussions related to implementing the Shifts in Classroom Practice *and supporting emergent multilinguals in developing the Mathematical Practices. While many are not specific to mathematics, the ideas can be discussed as they can be applied to mathematics teaching.*

TESOL International Association

http://www.tesol.org

This is a professional organization much like NCSM or NCTM, but it is focused specifically on the needs of emergent multilingual students. It offers a strong collection of books, hosts conferences, lists advocacy resources, and distributes news briefs on international and national issues. There is a lot here that is general and specific to mathematics.

TODOS: Mathematics for All

www.todos-math.org

TODOS is a highly energetic and engaged professional organization focused on advocating for strong mathematics for every child, but it also has a strong focus on Latino students. The website offers numerous resources for learning how to support multilingual students, as well as how to advocate for them. Joining the organization is reasonable and gives access to even more resources.

Understanding Language

http://ell.stanford.edu/content/supporting-ells-mathematics

This is a landing page for a resource that provides ideas for supporting language development. Beyond a more detailed overview of "best practices" with multilingual students, ideas and templates are provided to support student reading and vocabulary development. Sample lessons provide strong exemplars and can be useful in professional development.

NOTES

Coach's Toolkit

These tools are a menu from which you can select any that make sense for your setting/context. They can be used independently or as part of a coaching cycle. You may start with the self-assessment, which can guide you in deciding which of the other tools may be most useful. If using these tools for a coaching cycle, mix and match as you like or use one of the combinations we suggest in the diagrams that follow. The tools in this chapter include instructions to the coach and the teacher. You can download copies of the tools that only have instructions for the teacher at **resources.corwin.com/ mathematicscoaching**.

Self-Assess

9.1	Connecting *Shifts* to Supporting Emergent Multilingual Students Self-Assessment

Plan

9.2	Culturally Responsive Mathematics Instruction (CRMI)
9.3	Planning Strategies to Support Emergent Multilingual Students
9.4	Diagnostic Interview

Gather Data

9.5	Focus on Culturally Responsive Mathematics Instruction (CRMI)
9.6	Teaching to Support Emergent Multilingual Students

Reflect

9.7	Reflecting on Support for Emergent Multilingual Students
9.8	Reflecting on Using a Diagnostic Interview

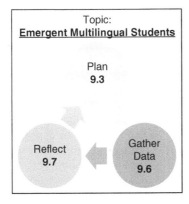

Topic:
Emergent Multilingual Students

Plan
9.3

Reflect **9.7** ← Gather Data **9.6**

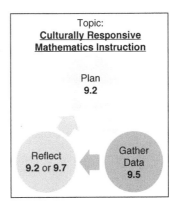

Topic:
Culturally Responsive Mathematics Instruction

Plan
9.2

Reflect **9.2 or 9.7** ← Gather Data **9.5**

Additional Tools in Other Chapters

See the many tools in Chapter 8 (Differentiating Instruction for All Learners) that can be used with a focus on emergent multilingual students. The tools in Chapter 6 (Formative Assessment) and Chapter 7 (Analyzing Student Work) can provides ways to gather data to better understand students' mathematical representations, explanations, and mathematical understandings.

 To download the coaching tools for Chapter 9 that only have instructions for the teacher, go to **resources.corwin.com/mathematicscoaching.**

9.1 Connecting *Shifts* to Supporting Emergent Multilingual Students Self-Assessment

Instructions to Coach: Ask teachers (individually or as part of a PLC activity) to self-assess where they position themselves on each of these Shifts in Classroom Practice related to content and worthwhile tasks. Use the following questions during a coaching conversation or in a PD setting to support teachers (and to help you decide which tools may be most useful from this chapter). A one-page version of this tool without this note is available for download.

Instructions: The *Shifts in Classroom Practice* listed below have specific connections to content knowledge and worthwhile tasks. Put an X on the continuum of each *Shift* to identify where you currently see your practice.

Tool 9.1 Shifts

Shift 1: From *stating-a-standard* toward *communicating expectations for learning*

| Teacher shares broad performance goals and/or those provided in standards or curriculum documents. | → | Teacher creates lesson-specific learning goals and communicates these goals at critical times within the lesson to ensure students understand the lesson's purpose and what is expected of them. |

Shift 2: From *routine tasks* toward *reasoning tasks*

| Teacher uses tasks involving recall of previously learned facts, rules, or definitions and provides students with specific strategies to follow. | → | Teacher uses tasks that lend themselves to multiple representations, strategies, or pathways encouraging student explanation (how) and justification (why/when) of solution strategies. |

Shift 3: From *teaching about representations* toward *teaching through representations*

| Teacher shows students how to create a representation (e.g., a graph or picture). | → | Teacher uses lesson goals to determine whether to highlight particular representations or to have students select a representation; in both cases, teacher provides opportunities for students to compare different representations and how they connect to key mathematical concepts. |

Shift 4: From *show-and-tell* toward *share-and-compare*

| Teacher has students share their answers. | → | Teacher creates a dynamic forum where students share, listen, honor, and critique each other's ideas to clarify and deepen mathematical understandings and language; teacher strategically invites participation in ways that facilitate mathematical connections. |

Shift 6: From *teaching so that students replicate procedures* toward *teaching so that students select efficient strategies*

Teacher approaches facts and procedures with the goal of speed and accuracy.	Teacher provides time for students to engage with mathematical problems, developing flexibility by encouraging student selection and use of efficient strategies; teacher provides opportunities for students to evaluate when a strategy is best suited for the problem at hand.

Tool 9.1 Reflection Questions

1. What do you notice, in general, about your self-assessment of these *Shifts in Classroom Practice*?

2. What might be specific teaching moves that align with where you placed yourself on the *Shifts*?

3. What might be specific teaching moves that align *to the right of* where you placed yourself on the *Shifts*?

4. What might be some professional learning opportunities to help you move to the right for one or more of these *Shifts*?

9.2 Culturally Responsive Mathematics Instruction (CRMI)

Instructions to the Coach: This topic applies to all students, including emergent multilingual students. This tool can be used as a task analysis and/or for planning a lesson for a coaching cycle. The same questions can be used in a reflecting conversation.

Instructions: Use the reflection questions to analyze a task and prepare a lesson that meets the needs of every student, including your emergent multilingual students.

CRMI	Reflection Questions to Guide Teaching and Assessing
The content is about important mathematics, and the tasks performed by students communicate high expectations.	• Does teaching focus on understanding big ideas in mathematics? • Are students expected to engage in problem-solving and generate their own approaches to problems? • Are connections made among mathematical representations? • Are students justifying their strategies and answers, and are they presenting their work?
The content is relevant.	• In what ways is the content related to familiar aspects of students' lives? • In what ways is prior knowledge elicited/reviewed so that all students can participate in the lesson? • To what extent are students asked to make connections between school mathematics and mathematics in their own lives? • How are student interests (events, issues, literature, or pop culture) used to build interest and mathematical meaning?
The instructional strategies develop positive mathematical identities.	• In what ways are students invited to include their own experiences within a lesson? • Are individual student approaches presented and showcased so that all students see their ideas as important to the teacher and their peers? • Are alternative algorithms shared as a point of excitement and pride (as appropriate)?
Each student's contributions are respected and valued.	• Are students invited and expected to engage in whole-class discussions in which they share ideas and respond to each other's ideas? • In what ways are roles assigned so that every student feels that he or she contributes to and learns from other members of the class? • How do I ensure that all students' contributions are valued by their peers?
Changes I will make to the task/handout/problem set to make the task more culturally responsive:	
Changes I will make to my instructional strategies to make the task more culturally responsive:	

Based on Van de Walle, J. A., Bay-Williams, J. M., Lovin, L. H., and Karp, K. S. (2018). Teaching Student-Centered Mathematics: Grades 6–8 *(3rd ed.). New York, NY: Pearson.*

9.3 Planning Strategies to Support Emergent Multilingual Students

Instructions to the Coach: Beyond using this tool in a coaching cycle or for lesson study, you can use it to analyze a video, assigning different strategies to different groups of teachers. Or distribute a task and have teachers consider these strategies as they plan how to implement the task.

Instructions: The 10 statements in the table are research-based strategies for ensuring emergent multilingual students have access and agency in learning mathematics. Record ideas for how you might incorporate these strategies in your lesson/unit.

Strategy to Support Emergent Multilinguals	*Specific Ideas for a Lesson or Unit*
1. Communicate high expectations.	
2. Make content relevant.	
3. Establish norms for participation.	
4. Honor native language.	
5. Set content and language goals.	
6. Provide vocabulary support.	
7. Use comprehensible input.	
8. Use cooperative groups strategically.	
9. Select tasks with multiple entry and exit points.	
10. Use diagnostic assessment tools.	

9.4 Diagnostic Interview

Instructions to the Coach: Use this template to co-design a diagnostic interview. Data gathering can be done by the teacher (via the interview), or you can observe the interview and take notes on the strategies used by the teacher to elicit student thinking.

Instructions: Select or design one to three tasks for the student to demonstrate and/or explain a concept. (The interview should take no more than 20 minutes, so the number of items relates to how long it will take for the student to solve each.) The National Assessment of Educational Progress (NAEP) is a great place to find interview items, as the Questions tool allows you to search for items by content, age, item type, and so on. Consider tasks on the same mathematics concept, but where one may be a story problem, another may just be numbers/symbols, and another may be very visual, with fewer words and fewer symbols.

Content to be assessed:

Task	*Interview Probes*
1.	
2.	
3.	

Interview Recommendations

1. If you don't know the student, begin with introductions. Smile—it is important you seem approachable and interested.

2. Explain that you are interested in how he or she is thinking and will therefore be asking questions about how she or he is solving the problems.

3. Pose the first problem. Watch the student solve the task; give sufficient wait time.
   - If student is stuck, ask what he or she does not understand—is it language related? Is it mathematics related? Pose the question a different way, draw a picture, or remind the student of something that might help him or her get started.
   - If the student solves the problem correctly, ask for an explanation of what he or she thought the problem was asking and how he or she solved it. In many cultures, mental math is highly valued, so do not assume that if you don't see much written down, the student is stuck or guessing.
   - Repeat for other problems. Keep interview short (5–15 minutes, depending on age).

4. Thank the students for sharing her or his thinking with you.

 Available for download at **resources.corwin.com/mathematicscoaching.** Copyright © 2018 by Corwin.

9.5 Focus on Culturally Responsive Mathematics Instruction (CRMI)

Instructions to the Coach: Write observed examples in each category or record actions or observations on a separate page. During the reflecting conversation, place actions in each category.

Important Content and High Expectations	Positive Mathematical Identities
Content is grade-level appropriate and includes high-level thinking, decision-making, and reasoning.	Students share experiences, connect to their lives, use their own strategies, and use multiple ways to show understanding

Relevant Content	Each Student's Contributions Valued
The content is connected to student lives and has relevant contexts, multiple representations, and so on.	Justification is used to determine the correctness of solutions, equitable sharing strategies used, and choices provided.

9.6 Teaching to Support Emergent Multilingual Students

Instructions to the Coach: With the teacher, identify particular students to observe (this tool is set up for three, but it can be adapted). Record these students' actions and participation, along with teacher moves that are intended to support these students. After the observation, map the data to the categories in Tool 9.3.

Student's Name	Student's Actions (Movements)	Student's Responses (Speaking, Reading, Writing, Listening)	Teacher Moves Related to the Student(s)

9.7 Reflecting on Support for Emergent Multilingual Students

Instructions to the Coach: This tool can be used in professional learning, PLCs, or a coaching cycle.

Instructions: Based on data gathered in a lesson, reflect on the following questions.

1. To what extent did your emergent multilingual students understand the big ideas of the lesson?

2. In what ways do you think you connected the lesson to students' experiences and prior knowledge? In what ways did the task you selected connect to students?

3. How did you provide explicit opportunities to use native language, use English, and connect native language to English?

4. In what ways did your lesson support the development of positive mathematical identities?

5. How might you describe the balance of participation and the ways that students' contributions were treated by others?

6. To what extent did you engage your emergent multilingual students in using language—reading, writing, listening, speaking—throughout the lesson?

7. Did the ways you engaged emergent multilingual students in language support their learning of academic English without frustrating them or limiting their participation in the lesson?

8. In what ways did you provide attention to vocabulary, and how did this affect the lesson delivery and the emergent multilingual students' participation in the lesson?

9.8 Reflecting on Using a Diagnostic Interview

Instructions to the Coach: This tool can be adapted for use in professional learning, PLCs, or a coaching cycle.

Instructions: Based on your experience in implementing a diagnostic interview, reflect on the following questions.

1. Which of the items seemed to be the most difficult for the student?

2. In what ways did the wording of the items influence the ability of the student to solve the problem(s)?

3. In what ways did your rewording or adding visuals affect the student's ability to solve the problem?

4. Were problems with words or without words more challenging, and why do you think this might be the case for this student?

5. What are some things you learned through the diagnostic interview process ...
   a. about the student?

   b. about interviewing?

   c. about mathematics and/or language?

   d. about teaching?

6. In summary, what are you learning about diagnostic interviews that you want to remember and use in future teaching and assessing?

Previously published by Bay-Williams, J., McGatha, M., Kobett, B., and Wray, J. (2014). Mathematics Coaching: Resources and Tools for Coaches and Leaders, K–12. New York, NY: Pearson Education, Inc.

Chapter 10

Supporting Students With Special Needs

Dear Coach,

Here is support for you as you work with teachers on supporting students with special needs.

In the COACH'S DIGEST ...

Overview: Strategies for supporting both mathematical and language development for students with special needs to read and/or to download and share with teachers.

Coaching Considerations for Professional Learning: Ideas for how to support teachers in considering strategies to support students with special needs.

Coaching Lessons From the Field: Mathematics coaches and leaders share how they have supported students with special needs using ideas and tools in this chapter.

Connecting to the Framework: Specific ways to connect selected *Shifts* and Mathematical Practices to supporting students with special needs.

Coaching Questions for Discussion: Menu of prompts for professional learning or one-on-one coaching related to teaching students with special needs.

Where to Learn More: Articles, books, and online resources for you and your teachers to learn more on this topic!

In the COACH'S TOOLKIT ...

Seven tools focused on equitable mathematics teaching, for professional learning or coaching cycles.

Coach's Digest

In the Coach's Digest, we begin with an overview of supporting students with special needs, written to teachers (and to you, the coach). As you read the Overview, the following questions might help you consider it in terms of your role as a coach:

- If I build on my teachers' strengths, as I am asking them to do with their students, where might I start in implementing these strategies?
- What types of professional learning activities, readings, and/or coaching cycles/lesson study might support their efforts to implement these ideas?
- What data might we collect to see if the strategies are indeed supporting students with special needs?

Overview of Supporting Students With Special Needs

A review of research on what classroom practices most impact student conceptual understanding found two strategies: making mathematical relationships explicit and engaging students in productive struggle (Hiebert & Grouws, 2007). In addition, *Adding It Up* (National Research Council, 2001), a review of research-based practices, suggests that students are best served when attention is given to (1) connecting and organizing knowledge around big conceptual ideas, (2) building new learning based on what students already know, and (3) incorporating students' informal knowledge of mathematics. We begin this brief discussion with these important findings for all students because too often strategies for students who have special needs are the opposite of these ideas—breaking information into minute pieces such that the big idea is completely absent from the lesson, or avoiding story problems, which have the potential to connect to a student's prior knowledge as well as to the underlying concepts. In many cases, the modifications intended to make the mathematics easier play to students' weaknesses, not their strengths.

Mathematics learning disabilities are cognitive differences, not cognitive deficits (Lewis, 2014). In particular, students with special needs struggle with forming mental representations of mathematical concepts, keeping numbers in working memory, organizing themselves, self-regulation, and generalizing. Therefore, instructional strategies that support students with special needs must address these challenges while still teaching mathematics that engages students and has them thinking at a high level. Research provides strong evidence of strategies that can help students overcome their processing challenges and successfully learn mathematics (Gersten et al., 2009; NCTM, 2007). Four are briefly described here.

Explicit Strategy Instruction

Explicit strategy instruction is not the same as direct instruction, and it is not telling students how to do a procedure. The middle name of this strategy is key—*strategy* instruction. Students with special needs benefit from hearing different strategies, practicing different strategies, and having guidance on how to go about selecting a strategy. The key is that concepts, processes, and connections are made more explicit. Rather than demonstrate a strategy, you are trying to *make the decision-making visible*. Teacher-led explanations of concepts and strategies help students to make connections between new knowledge and concepts they already know. Consider the collection of problem-solving strategies (e.g., draw a picture, make a list, think of a simpler problem, etc.). Explicit strategy instruction means illustrating what each of these looks like and practicing it before moving on to having students select a strategy. Concrete models support explicit strategy instruction. But rather than have unstructured or exploratory time, the teacher might model how to use the manipulatives and then give the students a new problem to explore using the tool in the same way.

Think-Alouds

Think-alouds make thinking processes explicit, so it is not surprising that think-alouds help students to learn strategies, processes, and how to do mathematics. Using think-alouds effectively requires taking time to consider what students know and what they might think to do, and talking aloud about how they might think through the problem. This is particularly important in solving open-ended tasks and high-level thinking tasks. Assume, for example, that rather than have students compute the cost of 5 adult tickets and 3 child tickets to the movies ($9 and $5 respectively), the teacher wants to open up the task, asking how many people might be able to go to the movies if they have $100. A think-aloud might begin with this:

Let's see, I have $100, and I can take some adults and some children to the movies. I don't have very much information. Maybe I could just try something first with friendly numbers.

Then, after trying this a teacher might think this aloud:

I need to organize this information so I can keep track of my trials.

Eventually, roles reverse as the student thinks aloud, helping the student to process information and solve worthwhile tasks and helping the teacher to formatively assess.

Concrete–Representational–Abstract (CRA)

The **concrete–representational–abstract (CRA)** instructional process is also often called *CSA*, with the *S* standing for *semi-concrete*, but *representational* is very fitting for mathematics where representations of concepts help students to develop abstract concepts. Either way, the idea is that teaching begins with ideas that are concrete, visible, and familiar, gradually moving toward mathematics concepts. In a combination with explicit strategy instruction, this approach benefits students with disabilities (Flores, Hinton, & Strozier, 2014; Mancl, Miller, & Kennedy, 2012). It is a mistake to think of CRA as a linear progression, however, because movement toward abstract ideas should continue to be connected to concrete ideas (manipulatives and contexts), as well as related representations (number lines, area models, sketches, graphs, etc.).

Peer-Assisted Learning

Because of the need to have ideas explicit, to think aloud, and to make connections between concrete and abstract ideas, working with peers is very helpful. Students with special needs benefit from seeing how other students model and explain mathematics (McMaster & Fuchs, 2016). **Peer-assisted learning** and support provides that just-in-time assistance. Therefore, it is helpful to pair students with special needs with peers who are able and willing to use think-aloud and/or to be explicit about their strategy.

NOTES

_____

_____

_____

_____

_____

_____

_____

_____

_____

_____

_____

_____

_____

_____

Coaching Considerations for Professional Learning

The suggestions here add to what has already been shared in Chapter 8 for supporting teachers as they work to better support every student.

1. ***Organize time for co-teachers to plan.*** Having the support of experts to assist students with special needs is important to those students' success. But most special education teachers are not mathematics specialists and therefore are less knowledgeable about the big ideas, mathematics learning progressions, and Mathematical Practices. On the other hand, they are likely more knowledgeable in strategies such as the ones suggested earlier. Try to include the special education teacher in professional development that focuses on both mathematics learning and the previous strategies. Imagine co-teaching pairs solving a worthwhile task and then practicing think-aloud on each other through a role-play scenario. Consider having the pair co-plan a lesson using Tool 10.2 or 10.3. Other options include Tool 8.3 (Planning for Diverse Learners) and Tool 3.6 (Worthwhile Task Analysis). Also, you can provide mathematics-specific readings on working with students who have special needs for the co-teaching team to read and discuss (see the Where to Learn More section).

2. ***Distinguish between explicit strategy instruction and direct instruction.*** Explicit strategy instruction, though teacher-led, is not the same as direct instruction. In fact, listening to a lecture or explanation about a concept does not work well for students who have processing, memory, and attention challenges. Explicit strategy instruction involves helping students focus on key ideas that might be implicitly illustrated with a manipulative or example, but that must be made explicit for a student with a learning disability. Explicit strategy instruction means being intentional about drawing attention to important mathematical relationships and connections—it is much more engaging, concrete, and student-centered than typical direct instruction. For questioning, see Tool 5.2, which focuses on high-level questioning, anticipating student challenges, and layering in additional questions to address these challenges (see also Tools 10.2, 10.4, and 10.6, which focus on explicit strategy instruction).

3. ***Practice adapting tasks.*** Review the ways to differentiate a task (see Chapter 8 Overview and Tool 8.4). One way to modify a task is to increase the structure but preserve high-level thinking, rather than remove the thinking opportunities (see Tools 3.5 and 3.6). Help teachers remember that students who have processing challenges still have high intelligence—they don't need easier problems; they simply need the problems adapted so that they can think more easily about them. In other words, modifying does not mean simplifying! A key focus area is procedural fluency (see Chapter 3). Too often, students with special needs are subjected to memorizing and practicing procedures they do not understand. Use Tool 3.3 to help teachers consider what it means to be fluent for a particular procedure and then to consider how to use explicit strategy instruction, think-alouds, and so forth to help students learn to select efficient strategies (see Tools 10.2, 10.3, 10.4).

4. ***Discuss how to limit student frustration and build perseverance.*** When students struggle, teachers want to remove that struggle. This is particularly true with students who have special needs because they may be unwilling to try if the task looks too daunting. Key to helping teachers really support the learning of students with special needs is to focus on whether the helping strategy removes the cognitive challenge or not. For example, removing story problems does not help a student learn to solve problems, but teachers must spend extra time building meaning for the story. This may involve removing the numbers (so students don't try to jump into some mathematics before they understand the problem). The "Three

Read Protocol" (an online search will lead to several great sites and examples) is one such strategy. Beyond reading comprehension, using mnemonics, drawing explicit attention to the meaning of symbols and terminology, using friendlier numbers, and adjusting visual displays (of a handout or projection, removing distracting information) are effective ways to remove barriers and build perseverance (see Tools 10.2 and 10.5).

5. *Focus on engagement.* Because students with special needs require more processing time, an answer is often given before they have even begun to think about the question. Wait time is critical. So are opportunities to respond to the questions being asked! The Chapter 4 Overview and the tools in Chapter 4 can be highly effective in making sure students with special needs are full participants in the lesson (and thereby getting more opportunities to make connections and practice).

Coaching Lessons From the Field

I have noticed three themes as I work with teachers in trying to make tasks accessible to a student with special needs while keeping the cognitive challenge. First is the tendency to remove the challenge of a task instead of adding structures. So we focus on scaffolds and other ways they can provide more structure in a lesson but still have opportunities for high-level thinking (like using the think-aloud and the Three Read Protocol). Second, a focus on manipulatives is really important. Teachers see the value in the concrete representation, but we need to really work on connecting the manipulative to the mathematical ideas. I have teachers complete the PICS Page (Tool 3.4) so they can think about the knowledge for a topic and then consider how they will make those aspects visible and connected. That is where graphic organizers and foldables can be really useful, and we talk about how to pair these with questions to help students make connections. Third, I notice that as tasks are opened up and as tools are used, teachers are impressed with the thinking of their students ... and that is a sign we have really gotten to a new place in supporting students with special needs with long-term mathematics success in mind.

—Melisa Hancock, Mathematics Consultant
Gail, TX

Connecting to the Leading for Mathematical Proficiency (LMP) Framework

As teachers focus on supporting students with special needs, it is important for them to make explicit connections to the *Shifts in Classroom Practice* and the Mathematical Practices. The brief paragraphs that follow provide ideas for making these connections. You and your teachers can continue to add to these ideas (see also Tool 10.1 for connecting to the *Shifts*).

Connecting Supporting Students With Special Needs to *Shifts in Classroom Practice*

Shifts	1	2	3	4	5	6	7	8

Shift 2: From *routine tasks* toward *reasoning tasks*

Teacher uses tasks involving recall of previously learned facts, rules, or definitions and provides students with specific strategies to follow. ⟶ Teacher uses tasks that lend themselves to multiple representations, strategies, or pathways encouraging student explanation (how) and justification (why/when) of solution strategies.

All students benefit from teaching that focuses on using multiple representations, strategies, and pathways, complete with complex tasks that require reasoning. Students with special needs, however, need to have structures in place to help them select strategies. As described in the Overview, that might include direct modeling and practicing of strategies or representations, then adding think-alouds and other explicit strategy instruction on when to choose which strategy.

Shift 3: From *teaching about representations* toward *teaching through representations*

Teacher shows students how to create a representation (e.g., a graph or picture). ⟶ Teacher uses lesson goals to determine whether to highlight particular representations or to have students select a representation; in both cases, teacher provides opportunities for students to compare different representations and how they connect to key mathematical concepts.

Teachers of students with special needs actually need to instruct *across* this continuum, beginning with showing representations and giving students opportunities to practice them. But instruction must continue so that the representation is connected to mathematical ideas (CRA). Students must get to a place where they select a representation—for example, using think-alouds or generating a priority list ("First, I will see if ____ strategy will work. If it is not a good fit, my second choice is ____ ."). As students compare representations, teachers must use talk moves (see Chapter 5, Figure 5.2, and Tools 5.3 and 5.8) to ensure that students with special needs are making the connections.

Shift 6: From *teaching so that students replicate procedures* toward *teaching so that students select efficient strategies*

Teacher approaches facts and procedures with the goal of speed and accuracy. ⟶ Teacher provides time for students to engage with mathematical problems, developing flexibility by encouraging student selection and use of efficient strategies; teacher provides opportunities for students to evaluate when a strategy is best suited for the problem at hand.

Students with special needs can fall victim to practices that are ineffective, such as memorizing. Further, tedious practice of meaningless procedures makes mathematics learning distasteful at best. Tool 3.3 can help broaden the focus of learning goals related to procedures for all students.

Shift 7: From *mathematics-made-easy* toward *mathematics-takes-time*

Teacher presents mathematics in small chunks so that students reach solutions quickly.

→ Teacher questions, encourages, provides time, and explicitly states the value of grappling with mathematical tasks, making multiple attempts, and learning from mistakes.

This *Shift* is absolutely a priority when working with students with special needs. Learning important mathematics takes time! And students with special needs require additional explicit strategy instruction plus opportunities to be able to practice these strategies and eventually get to where they can select strategies. All students benefit from a classroom where multiple attempts and errors are viewed as part of doing mathematics. Students with special needs particularly benefit from comparing examples and non-examples and from analyzing worked examples that have been done incorrectly.

> → Certainly, the *Shifts* provide many classroom practices that support all learners, including students with special needs. Students with special needs benefit from being reminded of what they have learned and having the opportunity to make explicit connections between previously learned and new content. This helps students build understanding and positions them to be able to explain what they are learning. Students with special needs need to know that effort matters. These four *Shifts* make mathematics accessible to students who have special needs. See Tools 10.1, 10.4, and 10.6, all of which focus on effective teaching specific to students with special needs.

Connecting Supporting Students With Special Needs to Mathematical Practices

MPs **1** **2** 3 **4** **5** 6 **7** 8

As noted in Chapters 8 and 9, meeting the needs of all learners permeates all of the Standards for Mathematical Practice. Mathematical Practices are critical proficiencies for students who have special needs as a means for them to learn the mathematics content; they are not just another thing students need to learn.

1. *Make sense of problems and persevere in solving them.* Because of the processing challenges that many students with special needs have, this particular Mathematical Practice is of critical importance. Students who have special needs have difficulty identifying entry points to a problem. Specific strategies must be put into action that do not diminish the challenge of a task, but instead provide increased structure. Increased structures are needed to make explicit the connections among problems and for students. Providing these additional structures increases student success, builds student confidence, and therefore affects their willingness to persevere in solving problems. If teachers use Mathematical Practice 1 as a goal for students with special needs, then the focus of instruction will be on developing mathematical proficiency, a much more ambitious and necessary goal that includes, but goes well beyond, skill acquisition.

2. ***Reason abstractly and quantitatively.*** Students need to learn to analyze a situation—a challenge for those that have difficulty comprehending abstract ideas and attending to relevant aspects of a problem. Instruction can be adapted to include more explicit attention to the processes needed. Note that this Practice states that students not only can generalize, but they can also come up with an example to help them reason. Coming up with an example (something concrete or representational) is an excellent strategy to support students with special needs to reason abstractly.

4. ***Model with mathematics.*** Students need to learn to analyze a situation and determine how to represent it abstractly—both are challenges for students who have special needs. Making explicit connections between words and symbols (and back again) can help to see what phrases indicate certain operations. This is *not* a keyword approach, but a reading-for-comprehension approach (What in the story told you that it was a multiplication situation? Could you have written it as division? How?).

5. ***Use appropriate tools strategically.*** Students who have special needs benefit from concrete tools (CRA model). Key to this Mathematical Practice is being able to choose tools, yet the act of choosing can be too open-ended for some students with special needs. This can be scaffolded by first assigning tools, then limiting choices and explicitly discussing choices, and eventually asking students to make their own choice of tools.

7. ***Look for and make use of structure.*** Noticing structure can help in many areas of mathematics, yet generalizing is particularly challenging for students who have special needs. Thus, instruction must provide explicit prompts to help students focus on patterns they should be noticing. For example, if a task asks students to factor $x^4 - 16$, a teacher might ask students to tell what they notice about these three equations and explain why they are true:

$$2^8 = (2^4)^2 \quad (10^4) = (10^2)^2 \quad x^6 = (x^3)^2$$

Then, ask students how that pattern can be applied to x^4 and to *16* in order to rewrite the expression as the difference of two squares.

Questions Related to the Focus Zone: Supporting Students With Special Needs

In addition to the questions in Chapter 8 on meeting the needs of all learners, these questions can engage teachers in thinking about supporting students who have special needs:

1. What are the key mathematical ideas of the lesson? What strategies might you use to make the key ideas and connections explicit?

2. What are the specific learning needs of each student who has special needs? How will this be taken into consideration in the lesson? What strengths does each student bring to learning mathematics? How might these resources be accessed in the lesson?

3. What prior knowledge, interests, or activities does the student have that could increase his or her interest in and commitment to the lesson?

4. What aspects of this lesson might be the most challenging for this student (or students) given her or his special needs?

5. How might you incorporate tools to assist students in abstract reasoning (e.g., what manipulatives, visuals, stories, or recording tables)?

6. What are students asked to do to demonstrate understanding? In what ways can this be adapted to address learning needs (e.g., providing structured recording sheets, manipulatives, sentence starters, etc.)?

7. As the lesson is enacted, what actions might be taken in the introduction to ensure everyone understands the big idea of the lesson and what the problem is asking students to do? (How might you connect to prior learning?)

8. What might you do to ensure that the mathematical relationships are explicitly connected?

9. What might you do to encourage the student to persevere with the task/problem without taking away her or his thinking/reasoning?

10. In what ways might you support student ability to say or write strategies or solutions?

Questions Related to the LMP Framework

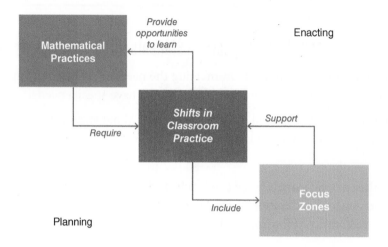

1. As you enact the ideas of supporting students with special needs, what do you see as related *Shifts in Classroom Practice*?

2. As you enact the ideas of supporting students with special needs, what do you see as opportunities to learn and, in particular, to develop the Mathematical Practices?

NOTES

_____

_____

_____

_____

_____

_____

_____

_____

_____

_____

_____

_____

_____

_____

_____

_____

_____

_____

_____

_____

_____

Where to Learn More

Books

Fennell, F. S. (Ed.). (2011). *Achieving Fluency in Special Education and Mathematics*. Reston, VA: NCTM.

> *This is an excellent collection of chapters focused on instructional strategies for teaching mathematics to students with disabilities. Chapters detail work in each content strand, addressing instruction and assessment. The entire book could be used with teachers, or particular chapters could be selected for a group of teachers or individual teachers to read.*

Gersten, R., Beckmann, S., Clarke, B., Foegen, A., Marsh, L., Star, J. R., & Witzel, B. (2009). *Assisting Students Struggling With Mathematics: Response to Intervention (RtI) for Elementary and Middle Schools* (NCEE 2009–4060). Washington, DC: National Center for Education Evaluation and Regional Assistance, Institute of Education Sciences, U.S. Department of Education. Retrieved from http://ies.ed.gov/ncee/wwc/publications/practiceguides.

> *This report analyzes, discusses, and assesses the research on intervention and makes specific recommendations based on the research. It contains tools for carrying out the recommendations and provides explicit examples for implementation in Tier 1, 2, and 3 interventions for mathematics.*

Articles

Allsopp, D., Lovin, L., Green, G., & Savage-Davis, E. (2003). "Why Students With Special Needs Have Difficulty Learning Mathematics and What Teachers Can Do to Help." *Mathematics Teaching in the Middle School, 8*(6), 308–314.

> *This is an excellent article for all grade levels, K–12. It addresses four learning characteristics that make doing mathematics difficult for students who have special needs and instructional strategies that address these challenges, making mathematics attainable. The vignettes, based on real-life experiences, make this a great article for professional development.*

Hodges, R., Rose, R., & Hicks, A. (2012). "Interviews as RtI Tools." *Teaching Children Mathematics, 19*(1), 30–36.

> *These authors describe how to use interviews to identify a student's strengths and challenges and how to use this information to design instruction within an RtI framework.*

Witzel, B., & Allsopp, D. (2007). "Dynamic Concrete Instruction in an Inclusive Classroom." *Mathematics Teaching in the Middle School, 13*(4), 244–248.

> *Through discussion and the use of two vignettes, this article highlights the use of manipulative materials for students with disabilities. The vignettes address research-based strategies and, because they are vignettes, lend themselves to reading prior to or during a workshop and then discussing/debating the strategies used (or that could be used) to support student learning.*

Online Resources

Mathematics and Learning Disabilities (LD Online)

www.ldonline.org

> *This site offers a wealth of resources, including assessment tools, teaching strategies, readings, videos, podcasts, and interesting articles on mathematical disabilities.*

NOTES

Coach's Toolkit

These tools are a menu from which you can select any that make sense for your setting/context. They can be used independently or as part of a coaching cycle. You may start with the self-assessment, which can guide you in deciding which of the other tools may be most useful. If using these tools for a coaching cycle, mix and match as you like or use one of the combinations we suggest in the diagrams that follow. The tools in this chapter include instructions to the coach and the teacher. You can download copies of the tools that only have instructions for the teacher at **resources.corwin.com/mathematicscoaching.**

Self-Assess

| 10.1 | Connecting *Shifts* to Supporting Students With Special Needs Self-Assessment |

Gather Data

| 10.4 | Effective Teaching for Students With Special Needs |
| 10.5 | Implementing Support Structures for Students With Special Needs |

Plan

| 10.2 | Challenges and Support Structures |
| 10.3 | Structuring a Lesson to Support Students With Special Needs |

Reflect

| 10.6 | Reflecting on Effective Teaching for Students With Special Needs |
| 10.7 | Reflecting on Structuring a Lesson for Built-In Success |

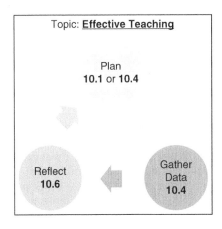

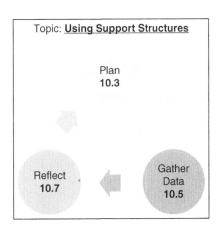

Additional Tools in Other Chapters

Since making connections is a major support structure for students with special needs, see Chapter 3 (Content and Worthwhile Tasks) for tools to help students build connections and develop a strong understanding of mathematics. Chapter 4 (Engaging Students) has many important tools for considering both how to engage and how to get all students to think at high levels. Many tools in Chapter 8 (Differentiated Instruction) can be used with a focus on students with special needs.

 To download the coaching tools for Chapter 10 that only have instructions for the teacher, go to **resources.corwin.com/mathematicscoaching.**

10.1 Connecting *Shifts* to Supporting Students With Special Needs Self-Assessment

Instructions to Coach: Ask teachers (individually or as part of a PLC activity) to self-assess where they position themselves on each of these Shifts in Classroom Practice *related to questioning and discourse. Use the questions that follow during a coaching conversation or in a PD setting to support teachers (and to help you decide which tools from this chapter may be most useful). A one-page version of this tool without this note is available for download.*

Instructions: The *Shifts in Classroom Practice* listed below have specific connections to questioning and discourse. Put an *X* on the continuum of each *Shift* to identify where you currently see your practice.

Tool 10.1 Shifts

Shift 2: From *routine tasks* toward *reasoning tasks*

Teacher uses tasks involving recall of previously learned facts, rules, or definitions and provides students with specific strategies to follow.	Teacher uses tasks that lend themselves to multiple representations, strategies, or pathways encouraging student explanation (how) and justification (why/when) of solution strategies.

Shift 3: From *teaching about representations* toward *teaching through representations*

Teacher shows students how to create a representation (e.g., a graph or picture).	Teacher uses lesson goals to determine whether to highlight particular representations or to have students select a representation; in both cases, teacher provides opportunities for students to compare different representations and how they connect to key mathematical concepts.

Shift 6: From *teaching so that students replicate procedures* toward *teaching so that students select efficient strategies*

Teacher approaches facts and procedures with the goal of speed and accuracy.	Teacher provides time for students to engage with mathematical problems, developing flexibility by encouraging student selection and use of efficient strategies; teacher provides opportunities for students to evaluate when a strategy is best suited for the problem at hand.

Shift 7: From *mathematics-made-easy* toward *mathematics-takes-time*

Teacher presents mathematics in small chunks so that students reach solutions quickly.	Teacher questions, encourages, provides time, and explicitly states the value of grappling with mathematical tasks, making multiple attempts, and learning from mistakes.

Tool 10.1 Reflection Questions

1. What do you notice, in general, about your self-assessment of these *Shifts in Classroom Practice*?

2. What might be specific teaching moves that align with where you placed yourself on the *Shifts*?

3. What might be specific teaching moves that align *to the right of* where you placed yourself on the *Shifts*?

4. What might be some professional learning opportunities to help you move to the right for one or more of these *Shifts*?

10.2 Challenges and Support Structures

Instructions to Coach: This tool can be used in professional learning to discuss ideas in general or used in a planning conversation or lesson study meeting connected to a specific lesson.

Instructions: Discuss ways in which to address specific challenges and incorporate each support structure for a lesson.

Challenges	How I Address/Remove These Challenges
1. Memory	
2. Attention	
3. Expressing ideas (verbally or in writing)	
4. Auditory, visual, or written perception	
5. Comprehension of abstract ideas	
Additional challenge(s):	

Support Structures	How I Add/Enhance These Structures
1. Give clear instructions and expectations.	
2. Provide visual displays and handouts that help organize thinking without having too much information.	
3. Organize writing (and speaking) about processes and mathematical concepts.	
4. Explicitly connect ideas (to previous learning, to familiar examples, and between concrete and abstract).	
5. Include self-reflection, self-assessment, and self-monitoring.	
6. Use support staff, including co-teaching.	
Additional support structure(s):	

Previously published by Bay-Williams, J., McGatha, M., Kobett, B., and Wray, J. (2014). Mathematics Coaching: Resources and Tools for Coaches and Leaders, K–12. New York, NY: Pearson Education, Inc.

10.3 Structuring a Lesson to Support Students With Special Needs

Instructions to Coach: This tool can be used in professional learning to discuss ideas in general or used in a planning conversation or lesson study meeting connected to a specific lesson. Use the same tool for data gathering, but in data gathering, you are the one recording notes.

Instructions: The lists (adapted from Van de Walle, Karp, & Bay-Williams, 2019) offer research-based strategies that support students with special needs. Identify one focus from each key area—or a particular area but multiple considerations—and brainstorm what ways that might be accomplished for a particular lesson.

Key Area	Planning Considerations What plans do you have for each student who has special needs in regard to ...	Notes
Organization and environment	➢ Student location (close to instruction) ➢ Reduce competing stimuli (noises, sights, or other distractions) ➢ Smooth transitions (from one phase of a lesson to the next) ➢ Post (and state) big idea of lesson	
Introducing the lesson (Engage)	➢ Build on prior knowledge ➢ Use a variety of visuals and concrete examples ➢ Vocabulary support ➢ Use friendly numbers (but same rigor) ➢ Clear directions (one direction at a time, check for understanding) ➢ Vary task size (so it is not overwhelming)	
Developing the lesson (Explore)	➢ Provide ways to organize work (e.g., graphic organizer, use of heuristics such as Polya's four-step process for problem-solving) ➢ Provide support in communicating ideas (writing and speaking) ➢ Emphasize big ideas ➢ Make mathematical connections explicit ➢ Encourage self-monitoring, self-assessment, and reflection	
Summarizing the lesson (Explain)	➢ Provide support in communicating ideas (writing and speaking), such as writing prompts ➢ Emphasize big ideas ➢ Emphasize mathematical connections among ideas ➢ Provide/solicit examples and non-examples ➢ Provide additional practice ➢ Offer strategies to help remember, as appropriate (e.g., mnemonics)	

10.4 Effective Teaching for Students With Special Needs

Instructions to the Coach: Have teacher(s) read the Overview section. For a planning conversation, a teacher can read this in advance and you can talk about how to attach these ideas to the lesson. Then, use this template to record data during the lesson. In a reflecting conversation, map the teacher moves to the four effective strategies and discuss the impact the moves had on the teacher's students with special needs.

Effective Teaching for Students With Special Needs	Teacher Moves	Student(s) Responses to Efforts (What Student Did, Said, Wrote)
Explicit instruction		
Think-alouds		
Concrete–representational–abstract (CRA)		
Peer-assisted learning		

10.5 Implementing Support Structures for Students With Special Needs

Instructions to the Coach: Use this with Tool 10.3, recording evidence (teacher statements, movements). Rather than you placing the moves within each structure, simply record moves as they occur in the lesson; then in the reflecting conversation, map the moves to the support structures.

Support Structures	Teacher Moves
➤ Give clear instructions and expectations.	
➤ Provide visual displays and handouts that help organize thinking without having too much information.	
➤ Organize writing (and speaking) about processes and mathematical concepts.	
➤ Explicitly connect ideas (to previous learning, to familiar examples, and between concrete and abstract).	
➤ Include self-reflection, self-assessment, and self-monitoring.	
➤ Use support staff, including co-teaching.	
➤ Other:	
➤ Other:	

10.6 Reflecting on Effective Teaching for Students With Special Needs

Instructions to the Coach: This tool can be used in professional learning, PLCs, or a coaching cycle.

Instructions: Reflect on your lesson or unit, review the data you have gathered from that lesson and/or unit, and identify successes and next steps for each of the effective teaching strategies.

Connection	My Success in Implementing This Element of Effective Teaching	Things to Try Next
Explicit instruction		
Think-alouds		
Concrete–representational–abstract (CRA)		
Peer-assisted learning		

1. What specific strategies, protocols, or routines have been particularly effective?
2. What are you noticing in terms of the use of these strategies and . . .
   a. students' participation?
   b. students' mathematical understanding?
   c. students' making connections and retention?

10.7 Reflecting on Structuring a Lesson for Built-In Success

Instructions to the Coach: This tool can be adapted for use in professional learning, PLCs, or a coaching cycle.

Instructions: Using data from your lesson, highlight the considerations you addressed and identify successes and next steps for each of the effective teaching strategies below.

Key Area	Evidence From the Lesson	Reflections on How It Went and What to Do Next
Organization and environment		
Introducing the lesson (Engage)		
Developing the lesson (Explore)		
Summarizing the lesson (Explain)		

1. What evidence do you see in the data that students with special needs understood the big ideas of the lesson?

2. What evidence do you see in the data that the strategies you implemented were successful (or not successful)?

Part III

Navigating a Successful Journey
Strategies and Tools for You, the Coach

If you have explored Chapters 1 through 10 of this book, you discovered lots of information and tools focused on the work you might do with teachers related to their practice. But you need more than information and tools to support teachers and teaching—this part of the book is for you as you focus on *your* practice! Navigating your journey as a coach involves having strong communication skills, the focus of Chapter 11. You may use this chapter for personal study or with other coaches to support each other in developing your own skills. We cannot overstate how important these skills are to all other aspects of your journey! In the final two chapters, Chapters 12 and 13, we take on what might be everything you need related to two common roles—coach as **presenter** (as in leading a professional development event) and coach as **facilitator** (as in facilitating professional learning communities). These three chapters, then, provide ideas and guidance to help you navigate all of your many interactions and activities as a mathematics coach.

Chapter 11

Interacting With Colleagues

Dear Coach,

Here is support for you to increase your effectiveness in interacting with colleagues.

In the **COACH'S DIGEST** ...

Overview: A brief look at three support functions of an effective coach and the importance of focusing on various ways of interacting with colleagues. Another aspect of interacting with colleagues focuses on specific coaching skills such as building trust, establishing rapport, listening, paraphrasing, and posing questions. For each skill, we provide this information:

- **Overview:** A review of the skill, offering at-a-glance needed information.

- **Tips:** Strategies for implementing the skill in your coaching.
- **Coaching Lessons From the Field:** Stories from mathematics coaches who have used these skills in their coaching.

Where to Learn More: Books for you on where to learn more about coaching skills!

Coach's Digest

This chapter is solely a Coach's Digest (so no tools in this chapter). As you read the Overview and engage in the activities describing each of five coaching skills, reflect on these questions:

- How intentional or aware am I of the support function in which I am engaging?
- What criteria do I need to consider when deciding which support function I need to engage in?
- To what extent do I automatically exhibit these coaching skills, or to what extent do I need to consciously remind myself to implement these skills?
- How might I get better at these coaching skills?
- How might I self-assess myself on the five coaching skills?

Overview of Interacting With Colleagues

As a coach, you wear many hats and are asked to complete many jobs. It may be difficult to really understand your role. Some districts have very specific job descriptions that help everyone in the district know the role of the coach; in other districts, the math coach is identified and placed into the job without guidance on how to focus her or his time. Costa and Garmston (2016) describe four specific **support functions** of school leaders that can help us in thinking about the role of a mathematics coach (see Figure 11.1).

Figure 11.1	Four Support Functions

Support Function	Description
Coaching	Nonjudgmental stance; supporting others to be more resourceful
Collaborating	Co-laboring; coming together to co-plan, co-teach, etc.
Consulting	Sharing expertise; modeling expert thinking
Evaluation	Making judgments about others, usually based on some sort of standard

Source: *Costa, A. L. & Garmston, R. J. (2016).* Cognitive Coaching: Developing Self-Directed Leaders and Learners. *Lanham, MA: Rowman & Littlefield.*

On which support function do you think you spend the most time? Let's briefly look at each of the support functions. First, you will notice that these support functions can apply to all coaching roles regardless of the title. In different districts and settings, coaches are called by a variety of titles: mathematics coach, mathematics specialist, instructional coach, lead teacher, resource teacher, and the list continues. Any of the people with these titles can serve in these support functions.

Coaching is a nonjudgmental stance, supporting the other person to be more resourceful. This is usually the support function that coaches know the least about because they have not been supported with professional development on how to be a coach (more to come on that later in the chapter!). Collaborating is working together—coming to the table to accomplish a shared goal for which everyone brings knowledge to the task at hand. Consulting is sharing your expertise—this is when you get to tell others what to do. But this needs to be done carefully; in fact, you really should ask permission before sharing your expertise (for example, you might say, "Could I share a suggestion?"). Evaluation is making judgments about others, usually based on a set of standards. In the very best scenario, mathematics coaches should not serve as evaluators; that role should be left to administrators. If you are asked to evaluate, it can have a negative impact on the trust you are trying to build with the teachers with whom you work. That leaves coaching, collaborating, and consulting that could be part of your role.

Many coaches find themselves "living" in the consulting support function. While this can, at times, be an efficient role (quickly just tell someone what to do), it doesn't serve teachers (and their students) well when you stay in this support function. If your goal is to support teachers to be more self-directed, they need to have opportunities to think for themselves. If you are always telling them what they should do, it can build dependence and not self-directedness. Instead, try to move to the coaching support function more often so you can support teachers in being more resourceful—for example, by guiding them through conversations with purposeful questions. Then, when you really need to go to consulting, do it occasionally and return to coaching.

Some coaches feel ill-equipped to "live" in the coaching support function because they have no training on the needed skills. Coaching skills are critically important and sometimes overlooked. In many settings, administrators assume that coaches know how to coach because they were effective as mathematics teachers. Good teaching skills do not always equate to good coaching skills, but as with effective teaching, there are skills (e.g., questioning) you *learn* that lead to effective coaching.

In this chapter, we will focus on four coaching skills that are critical for all coaching interactions: building trust and rapport, listening, paraphrasing, and posing questions. For each of these skills, we present a synopsis that includes a description of the skill, tips for effectively implementing the skill, and insights from the field. We recognize these skills are complex, and becoming proficient in them is an ongoing process that develops over time. Fortunately, there are many resources available to you that describe these and other coaching skills in greater detail that can support you in integrating these skills into your coaching practice. In particular, *Cognitive Coaching: Developing Self-Directed Leaders and Learners* (Costa & Garmston, 2016) was an important resource for this section. For more information about Cognitive Coaching[SM], visit www.thinkingcollaborative.com. Here, we focus on providing a very brief overview of the coaching skills that can positively impact your effectiveness as a coach.

Building Trust

Nearly every book written about coaching discusses the need for building trust. At some level, this just seems so obvious, yet it is mentioned again and again. When you talk to coaches in the field, they also declare the importance of building trust. In many of our personal and professional relationships, trust is simply something we often take for granted. However, when a coach has to begin building trust from the ground up, you sometimes need to examine a little more closely what behaviors and actions are actually involved in making that happen.

Trust is not something that happens by accident or in the moment; it develops over time. Being intentional is the key to building trust with the teachers with whom you work. Recent research on trust in schools provides insight into the criteria people use when deciding whether they are going to trust another person. Some of these aspects of trust are integrity, competence, reliability, benevolence, and openness (Bryk & Schneider, 2002; Covey, 2008; Tschannen-Moran, 2014). Understanding each of these aspects of trust and assessing your proficiency in each will support your efforts at building trust with others. As a coach, you cannot make a teacher (or anyone else, for that matter) trust you. What you can do is act in a trustworthy manner (displaying the aspects of trust just mentioned) so that the teachers you work with will choose to trust you.

Tips for Building Trust

- Make it your top priority!
- Initially establishing trust can be operationalized in various ways, depending on your situation. It might mean you help with bus duty or gather resources for a lesson. Even if these activities are not officially part of your job description, they go a long way in establishing trust at the beginning of a professional coaching relationship.
- Find ways to demonstrate that you genuinely care about the teacher's professional life.
- Make appropriate connections to the teacher's personal life.

- Be honest. Dishonesty destroys trust.
- Maintain confidentiality. Nothing destroys trust like a confidence that is broken.
- Be transparent in communication, decision-making, and sharing information.
- Demonstrate reliability by meeting deadlines, keeping appointments, and following through on your commitments.
- Demonstrate competence by maintaining your content, pedagogical content, and coaching knowledge and skills. This does not mean that you must know it all. However, it does mean that you consistently exhibit a desire for lifelong learning while also honestly acknowledging the power of learning from mistakes.

Coaching Lessons From the Field: Building Trust

I work with all the mathematics teachers in my building. One teacher, Sonja, would not engage in mathematical discussions with me, was not receptive to me visiting her classroom, and did not actively participate in the math PLC meetings that I facilitated. I tried everything and was out of ideas about how to engage this teacher. I talked with another mathematics coach in our district and asked her for some advice. As she and I discussed the situation, I came to realize that Sonja probably did not trust me. This was actually a surprising realization. I assumed since Sonja and I had been colleagues for several years the trust would be there when I moved into a coaching position. However, from Sonja's perspective, the relationship changed when I became the mathematics coach. I decided I needed to back up and work on building a new level of trust with Sonja before attempting to "coach" her.

I knew Sonja's children played soccer and decided that was the avenue I would use to begin to establish a more trusting relationship. When I found opportunities to casually interact with Sonja, I focused the conversation on her children and their soccer accomplishments. Sonja seemed eager to talk about her children. As I continued to interact with Sonja over the next few weeks, we discovered other common areas of interest we were unaware of despite being colleagues for several years. The time I spent establishing a trusting foundation with Sonja paid off in the long run. Eventually, Sonja became more engaged in the math PLC meetings and finally even invited me to help her plan some lessons.

Although the strategy of making connections to a teacher's personal life worked with Sonja (see Coaching Lessons From the Field: Building Trust), it may not work with every teacher. As a coach, you have to determine what will work with each teacher with whom you interact. Building trust is a very individualized activity.

Establishing Rapport

Costa and Garmston (2016) describe **rapport** as "comfort with and confidence in someone during a specific interaction" (p. 100). Thus, rapport happens in the moment—as compared to trust, which develops over time. However, rapport and trust are inextricably linked. As you establish rapport during specific interactions, you are laying the foundation for a trusting relationship. You may have said of a teacher, "She has great rapport with her students." What does that really mean? How do you establish rapport with someone?

Establishing rapport is really about making both verbal and nonverbal connections with another person. The most common way we build rapport with one another in a conversation is by making eye contact. Other ways we build rapport and connect with others are by nodding, using open body posture, and making comments indicating agreement.

Recent research in the field of cognition and neuroscience has identified *mirror neurons* in the brain that indicate we are hardwired to respond to what others do in our environment (Goleman, 2006; Iacoboni, 2008). These mirror neurons help explain why mirroring or matching the behaviors (nonverbal and verbal) of another person draws us into rapport. This can involve mirroring body language (posture and gestures); voice volume, inflection, and pitch; language choices; and the pace and energy of a conversation.

Tips for Establishing Rapport

- Begin by focusing on one aspect of establishing rapport and build to adding others. For example, begin by just trying to match the posture of the teacher you are coaching. When you feel comfortable with this aspect of building rapport, add a second, such as matching the pace and energy of a conversation.
- Attempts at establishing rapport should be subtle and a natural part of the conversation. If not, the attempt can seem like mimicking or mocking and can be offensive.
- When focusing on posture, be aware that some people have sensitivities to their personal space. Thus, although leaning forward into a conversation indicates interest in what the teacher is saying, leaning too far in can violate personal space.
- Matching gestures can be tricky, especially if you do not normally gesture yourself. Watch to see whether the teacher uses gestures to emphasize particular ideas or concepts. If so, try to gesture in a similar way when discussing those topics.
- When trying to match language, be mindful of cultural and generational differences. You do not have to use the *same* language, but try to match the intent of the teacher's language.

The key to building rapport is that if you are doing it well, it should not be noticed.

Coaching Lessons From the Field: Building Rapport

I coach at three middle schools. In the fall, one of my mathematics teachers asked me to sit in on a parent–teacher conference in which she was expecting an angry parent. As the mother entered the room, I could tell right away that she was upset. Her body language was rigid, and she had a grimace on her face. We welcomed her into the room and asked her to have a seat. I had arranged the chairs in a small circle without a table so I could focus on building rapport with the parent. The mother started off by explaining how upset she was that her son's grade had dropped in mathematics during the second grading period. I let the teacher address this issue while I focused on mirroring the mother's posture. After the teacher's discussion of the student's grades, I leaned in slightly and asked the mother if the student's grades had suffered in any of his other classes. As she responded, I noticed how she used her hands when talking about his classes (gesturing with her left hand) and the increasing amount of time he was spending in basketball practice (gesturing with her right hand). I paraphrased her frustration and just slightly turned up each of my hands as I mentioned the classes and basketball. I also attempted to match her frustrated tone without offending her to let her know I heard her concern. She began nodding her head in response to my statement, and I could see some of the tenseness in her body relax just a bit. I continued to focus on building rapport as the conversation continued for several more minutes. Even when the teacher was talking and I was listening, I still focused on mirroring the mother's posture. By the end of the conversation, the mother was thanking us for helping her think through the situation with her son. It was amazing to me how paying attention to rapport really changed the feel of the conversation.

Listening

Listening is a critical coaching skill. The kind of listening that is important in coaching is not the same as the everyday listening we experience in this fast-paced world! The type of listening coaches need to use has been called "mindful listening" (Tschannen-Moran & Tschannen-Moran, 2010), "committed listening" (Kee, Anderson, Dearing, & Shuster, 2017), and "empathetic listening" (Covey, 2013). This intense level of listening is most famously defined by Covey as "listening and responding with both the heart and mind to understand the person's words, intent, and feelings" (p. 252). *Truly* listening to someone with the intention to understand provides opportunities to build rapport and lays the groundwork for establishing a trusting relationship. Since most of us do not engage in this type of listening on a daily basis, it requires deliberate practice to become proficient.

Listening this intently is difficult because you have to focus completely on the teacher, avoiding temptations to think about how this relates to your own experiences or to something that happened somewhere else. William Isaacs (1999) suggests that "we are always projecting our opinions and ideas, our prejudices, our background, our inclinations, our impulses; when they dominate, we hardly listen at all to what is being said" (p. 84). When you are interacting as a coach, you must consciously refrain from engaging in these unproductive listening behaviors. Costa and Garmston (2016) organize these unproductive listening behaviors into three categories: **autobiographical listening, inquisitive listening**, and **solution listening**. Each is briefly described here:

Autobiographical listening is the desire to interject your personal thoughts and ideas into a conversation. You can understand why this is a barrier to productive listening. As soon as our mind wanders and starts thinking about "our story" and what we want to say, we can no longer focus on truly listening to the other person.

Inquisitive listening is becoming overly curious about the details of the person's story that might not be important to the conversation. We can find our minds becoming distracted with thoughts of "what if …," "I wonder why …," and "what happened then?" In our day-to-day conversations, we find ourselves drawn into inquisitive listening in our desire to be "in the know" about everything! However, when listening as a coach, we need to focus only on the information relevant to helping us understand the other person's perspective.

Solution listening is jumping in to offer a solution for the other person's problem. Helping others solve problems is such a part of who we are as educators that it is difficult to not offer solutions. However, if our mind shifts to thinking of a solution to offer, we cannot be intentionally listening in such a way as to understand the other person's perspective.

All three of these unproductive patterns of listening interfere with our ability to listen with the intent to understand the other person. As you coach, set these ways of listening aside in order to completely tune in to the other person.

Tips for Becoming a Good Listener

- If you are easily distracted, find a space in which distractions are limited.
- If you have a lot on your mind and that is interfering with your focus, spend five minutes writing down, or mentally processing, what is on your mind so that your mind is clear and ready to focus on the speaker.
- Occasionally, and with the permission of the one you are coaching, audio record a conversation and revisit it to analyze your questions. Check to see whether you avoided the unproductive patterns of listening described earlier.
- Building rapport sends a message that you are listening. Do this with occasional nodding, positive facial expressions, and a relaxed posture.

- We can filter what we hear others say based on our personal biases, beliefs, and values. In order to understand the speaker, you may need to provide feedback in the form of **paraphrases** or clarifying questions to make sure you understand the speaker's perspective (see the following sections for more information on paraphrasing and questioning).
- Practice, practice, practice. Being a good listener is something you can practice in everyday conversations.

For each of the following vignettes, identify the unproductive listening pattern that should be avoided.

Coaching Lesson From the Field: Listening—Vignette 1

Coach: So how do you think your lesson on fractions went today?

Teacher: I was really pleased with the way the students were engaged. I think using the fraction circles helped the students to understand what it means to "find a common denominator."

Coach: So using the manipulatives supported the students' understanding. What were some things you did that contributed to the success of the lesson?

Teacher: I think planning ahead about how to use the fraction circles and having all the materials ready for each group of students played a role in the successful flow of the lesson.

Coach: I know what you mean. I always found it helpful to plan ahead and have all materials out and ready. That extra bit of planning always helped my lessons to go smoother. I even created math kits for each student that contained all the manipulatives we would be using for the week. That way students had exactly what they needed. It worked really well for me. So in terms of the lesson objective, what is your assessment of the students' understanding?

Which unproductive pattern of listening did this coach use?

Autobiographical   Inquisitive   Solution

Coaching Lesson From the Field: Listening—Vignette 2

Coach: What are you hoping for students to accomplish in this lesson?

Teacher: I want them to understand subtracting with regrouping.

Coach: What exactly do you want them to understand?

Teacher: I want them to know that regrouping in subtraction is connected to making equivalent representations that we did in our place value unit.

Coach: The connection between these two ideas is key for you. How might you know if students have made that connection?

Teacher: I'm going to have students use base 10 blocks to explore the tasks just like we did when making equivalent representations. As they are working, I will walk around and visit each table to assess whether students are using what they learned about equivalent representations. I will also have a recording sheet for them to record their work that includes a structure for showing the equivalent representation.

Coach: You will observe their work and provide a recording sheet. You know, I think the best way to monitor this would be to have students show their work with the base 10 blocks using the document camera. That way, all students can see exactly what they did with the blocks. Have you thought about trying that?

Which unproductive pattern of listening did this coach use?

Autobiographical   Inquisitive   Solution

Coaching Lesson From the Field: Listening—Vignette 3

Coach: What's on your mind?

Teacher: I am in charge of planning the Family Math Night and wondered if you could help me think through it.

Coach: Sure! What would you say is your goal for Family Math Night?

Teacher: We really want our parents to understand that our math curriculum is focused on supporting students in developing conceptual understanding of mathematics; it is not just about learning step-by-step procedures or skills in isolation.

Coach: So you want to help them see that procedures, skills, and conceptual understanding are all important. What might be some strategies you will use to foster that understanding for parents?

Teacher: I was thinking about having different stations set up where parents and students can be engaged in math activities form the curriculum so students can explain to their parents what they have learned. We could also have a teacher at each station to answer questions and oversee the activity.

Coach: This is so interesting—I have been asked to help with a similar event at my daughter's school. How are you planning to get parents involved? Have you figured out how the best way to advertise the event? What about food—are you having snacks available?

Which unproductive pattern of listening did this coach use?

Autobiographical   Inquisitive   Solution

Paraphrasing

Paraphrasing, like listening and establishing rapport, builds trust. When paraphrasing is done well, it sends a message that you are listening and trying to understand the other person. Paraphrasing goes hand in hand with listening. In fact, you cannot offer an effective paraphrase unless you really listen. Paraphrasing is also closely linked to building rapport; when you paraphrase, you build rapport by matching the other person's words and volume.

Paraphrasing is a powerful tool because it clarifies that you have heard and correctly interpreted what the other person is saying. In fact, Costa and Garmston (2016) suggest that "paraphrasing is one of the most valuable and least-used tools in human interaction" (p. 48). We have all been in conversations where we realize, in the middle of the conversation, that not everyone is talking about the same thing! This happens in faculty meetings, in professional learning community discussions, and even at the family dinner table. Paraphrasing helps to eliminate this problem because you are rephrasing the other person's words, clarifying to make sure that everyone is on the same page before continuing the conversation.

Lipton, Wellman, and Humbard (2003) suggest that paraphrasing is like a "gift" given to the speaker because it "reflects a speaker's thinking back to the speaker for further consideration" (p. 54). Taking this idea a little further, Ellison and Hayes (2009) state that "when people hear their own words reflected back to them in a rephrased manner, it is often the first time they attend to the meaning of their own thinking" (p. 86). In coaching, both of these benefits of paraphrasing are critical: clarifying your interpretation of the other person's words and giving the teacher an opportunity to examine his or her own thinking.

Tips for Effective Paraphrasing

- As noted in the previous section, listen with the intention to understand. If you are not fully attending to the other person, you will not be able to give an effective paraphrase.

- Make the paraphrase shorter than the original statement. Think "big idea" when crafting your paraphrases. You should be talking less than the person you are coaching.
- Offer a paraphrase before asking a question. By starting with a paraphrase, you create a safe and inviting environment.
- Avoid "parroting"—saying exactly the same thing as the other person—as this can come across as mimicking.
- Do not use the word *I* in your paraphrases. Always make the paraphrase about the other person. Instead of saying, "What I hear you saying is that you're frustrated with the pacing guide," just say, "You're frustrated with the pacing guide."
- Practice, practice, practice! You can practice paraphrasing in every conversation you have. You can also practice when watching a television show by paraphrasing what you hear an actor say.

Coaching Lessons From the Field: Paraphrasing

It took me a while to learn how to paraphrase effectively. I had a rocky start but stayed with it and finally saw the benefits. I first learned about paraphrasing from a coaching seminar. I decided to try it and see what happened. I had a planning conversation with Kathy, a third-grade teacher, shortly after the seminar and went in thinking about being intentional about paraphrasing. I began the conversation by asking about the goals for the lesson. When Kathy responded, I repeated what she said, and she looked at me kind of funny. I went on with the conversation and asked how she would collect evidence from students that would indicate they had met the goal. Again, I repeated what Kathy said, but before I could ask the next question, Kathy looked at me and said, "Why are you repeating everything I say?" I felt terrible! I apologized and finished the rest of the conversation without paraphrasing. I left that conversation thinking that paraphrasing was a waste of time because it really interrupted the flow of the conversation.

At the next district coaching meeting, I asked some of the other coaches who had attended the seminar if they had tried paraphrasing. We began sharing our experiences and looked back over our notes from the seminar. I learned that I did remember to paraphrase before asking a question but I did not remember some other suggestions for effective paraphrasing, such as always make the paraphrase shorter and capture the essence of the original statement. So repeating exactly what Kathy said was not the way to do it—no wonder she looked at me funny!

Now fast-forward about three months: I have continued to practice paraphrasing and have gotten better because people do not even notice I am paraphrasing. I find paraphrasing to be so beneficial in conversations because it helps me to make sure I understand the teacher's thinking. For instance, one day Sue was explaining how she attempted to use some differentiation strategies to meet the needs of all learners in her class. Here is how part of the conversation went:

Sue: I tried some differentiation strategies today with my fifth-period class, but it didn't go well with a few students.
Me: So the behavior of some students interfered with the strategies you tried.
Sue: No, it wasn't their behavior. I really think I just didn't have the correct strategies for those three students.
Me: Oh, OK. It was a matter of finding the right strategy. What were some indications that this wasn't working for these students?

Based on some of our previous conversations about this class, I assumed the problem had been student behavior. The paraphrase I offered allowed Sue to clarify the situation, and the conversation went in a totally different direction. Paraphrases are so useful for making sure everyone is "on the same page," and that is a time saver!

Posing Questions

Posing questions is something we do every day, yet, as Barkley (2010) says, "Asking questions is an art—an elegant art" (p. 109). When posing questions in a coaching situation, you need to be intentional in the way you phrase questions so that you support the other person's thinking while also building trust in the relationship. Posing questions is third in the process of productive discussions—listen, paraphrase, pose questions, and repeat. Wait time is also critical in this interaction pattern (Costa & Garmston, 2016; Kee et al., 2017; Woleck, 2010). Pausing to allow yourself and the coachee time to think sends a message that thinking is important.

The purpose of posing questions in a coaching situation is to move the other person forward in her or his thinking. Reiss (2015) suggests that well-crafted questions empower teachers "to think about possibilities, explore the unknown, elaborate on or expand their thinking" (p. 127). Similar to asking questions in the classroom, we want to focus on higher-order thinking when asking questions to extend teachers' thinking. Glaser (2014) presents a conversational dashboard on which she categorizes questions into three levels: transactional, positional, and transformational. Questions at the transformational level allow coaches to explore teachers' perspectives and create a sense of mutual success and high trust. Posing questions in this manner indicates to the teacher that (1) you value his or her thinking and (2) you do not enter the conversation with an agenda. These types of interactions build trust.

Here are some sample questions that invite teachers to explore their thinking. Look them over—the Tips for Posing Questions in the following section should be apparent in them.

- What might be some strategies you have tried before that were successful?
- What are some connections between this goal and the standards?
- What seems most useful in this situation?
- What might be some of your choices?
- In what ways might you sequence those ideas?
- What are some specific patterns or trends that you are noticing?
- How did the lesson compare to how you planned it?
- What do you think might have been going on for students during the lesson?
- What are some criteria you used to decide about using manipulatives in this lesson?
- How might you summarize that student's thinking on this task?

Tips for Posing Questions

- Use plurals in your questions. This indicates the possibility of more than one answer (strategies, outcomes, thoughts).
- Embed tentative language—words such as *might*, *seem*, and *some*—to open the conversation to additional possibilities.
- Ask open-ended questions to allow the coachee to express her or his thinking. Open-ended questions provide the coach with valuable information about the coachee's attitudes, beliefs, and values.
- Use verbs that will elicit higher-order thinking, such as *compare*, *predict*, *evaluate*, *summarize*, and *prioritize*.
- Embed positive presuppositions in your questions to indicate that you presume positive intentions as well as competence in your coachee.

- Use an "approachable voice" (Grinder, 1993) to signal inquiry instead of interrogation. Grinder characterizes this voice as having a rhythm and describes it as the voice of building relationships.
- Think ahead and plan questions that work in particular situations. However, you must remain flexible during a conversation to go where the coachee takes you.

Coaching Lessons From the Field: Posing Questions

Asking questions is something I do all the time as a coach, and I thought I was pretty good at it. One of my colleagues shared with me that she video recorded some of her conversations and then analyzed her questions and paraphrases. I decided to try this and focused on the questions I asked. Well, first of all, watching myself and analyzing my conversations was painful! Even though it was difficult at first, it was a really valuable professional development experience for me. In fact, I now ask some of my coaching colleagues to look at my conversations with me. Here are a few things I noticed:

When I start a question with "why," this sometimes puts the teacher on the defensive. I found ways to take "why" out of my questions. For example, instead of saying, "Why did you decide to start the lesson with that example?" I might say, "What are some criteria you used to decide how to start the lesson?"

One of my colleagues pointed out my questions did not really invite the teacher to explore multiple options. I didn't really understand this at first, but by analyzing my questions for this characteristic, I began to see a pattern. One way I learned to alter my questions to address this was to use plural instead of singular words. For example, instead

of saying, "What approach are you planning for introducing this activity?" I would say, "What are some approaches you might use to introduce this activity?" While this might seem like a subtle difference, the impact is huge. My first question made it seem as if there was one right approach to use and I expected the teacher to know it! In contrast, my revised question invited the teacher to think about multiple approaches that could work.

I have found video recording my conversations and inviting colleagues to join me in analyzing them a very useful experience. There have been some obstacles to overcome to make this happen. First, not every teacher wants to be involved in a video recorded conversation. That is OK. I don't have time to do this with every conversation anyway. I have found a few teachers that are willing to support me in this, and that works. Of course, the other major issue is finding time for this work with everything else I have to do! I decided to set a goal of video recording one conversation a month and setting aside a one-hour block of time to focus on the analysis with a colleague. I can make this work with my schedule. I try to not compromise this monthly event with other things that need to get done. I view it as part of my professional growth.

Where to Learn More

Cheliotes, L. G., & Reilly, M. F. (2010). *Coaching Conversations: Transforming Your School One Conversation at a Time*. Thousand Oaks, CA: Corwin.

> *The authors include an entire chapter focused on committed listening, which describes the three listening set-asides and offers a fourth, judgment or criticism. Self-assessment tools and implementation strategies are included.*

Costa, A. L., & Garmston, R. J. (2016). *Cognitive Coaching: Developing Self-Directed Leaders and Learners*. Lanham, MA: Rowman & Littlefield.

> *This book, which describes the foundations of Cognitive Coaching[SM], includes an entire chapter on building trust. Each of the other coaching skills from this chapter is described in more detail.*

Covey, S. R. (2013). *The 7 Habits of Highly Effective People: 25th Anniversary Edition*. New York, NY: Simon & Schuster.

> *In this iconic book, Steven Covey outlines seven habits of people who live and work at a high degree of effectiveness. Of particular interest to coaches is Habit 5: Seek to Understand, Then to Be Understood. In this chapter, Covey outlines the principles of empathetic listening.*

Glaser, J. E. (2014). *Conversational Intelligence: How Great Leaders Build Trust and Get Extraordinary Results*. New York, NY: Bibliomotion.

> *This book is based on new insights from neuroscience but presents them in an understandable way. The author presents a framework for conversations to help you distinguish between conversations that lead to trust and those that don't. Practical strategies and tips for conversations to build trust are provided.*

Rizzatto, M., & Donelli, D. (2014). *I Am Your Mirror: Mirror Neurons and Empathy*. Giaveno, Italy: Blossoming Books.

> *The authors explain in very clear language the science behind mirror neurons. They describe how mirror neurons allow us to connect with others on a deep level, including connections to everyday social situations.*

Tschannen-Moran, M. (2014). *Trust Matters: Leadership for Successful Schools* (2nd ed.). San Francisco, CA: Jossey-Bass.

> *This book not only presents research on the importance of trust in schools, it also provides practical advice and strategies to make it happen. The role of trust between teachers and students, teachers and administrators, and schools and families is addressed. Trust-building activities and questions will support your efforts in developing trusting relationships in your school.*

Chapter 12

Presenting Professional Development

Dear Coach,

Here is support for you as you work to increase your effectiveness in presenting professional development.

In the **COACH'S DIGEST** . . .

Overview: A brief review of the research on presenting professional development.

Tips: Strategies for planning, implementing, and evaluating professional development.

Coaching Lessons From the Field: Mathematics leaders share how they used a tool or idea from this chapter.

Where to Learn More: Articles, books, and online resources for you on where to learn more about professional development!

In the **COACH'S TOOLKIT** . . .

Eleven tools focused on planning or evaluating PD and sample PD activities ready for you to use!

Coach's Digest

In the Coach's Digest, we begin with an overview of presenting professional development to support you in one of the many roles in which you engage. As you read the Overview, the following questions might help you consider it in terms of your role as a coach:

- Which of the seven characteristics of effective professional development are evident in the professional development I provide?
- In what ways am I infusing the four components necessary for teachers to incorporate new learning into their practice into the professional development I provide?
- How do I involve teachers in deciding about professional development needs?
- What might be the most important criteria I use to evaluate professional development?

Overview of Presenting Professional Development

You already know that being a mathematics coach means that you wear many hats! The previous parts of this book (Chapters 1–10) focus on the work you might do with individual teachers, perhaps in coaching conversations or even in small groups. However, another job that many mathematics coaches engage in is providing professional development for groups of teachers. We use the term *professional development* to describe a scheduled event in which a group of teachers convene for focused and structured learning; as mentioned earlier in the book, we use the term *professional learning* to refer to what we hope will occur at that event. In short, professional development is the event, and professional learning is the outcome.

Professional Development Options

When planning for professional development, it is important to take into consideration the research about effective practices (see Tools 12.1, 12.2, 12.3, and 12.4). In *Effective Teacher Professional Development*, Darling-Hammond, Hyler, and Gardner (2017, pp. v–vi) reviewed 35 research studies that showed a positive connection between professional development, teacher practices, and student outcomes. They identified seven characteristics of effective teacher professional development widely shared across the studies:

1. Is content-focused
2. Incorporates active learning
3. Supports collaboration
4. Uses models of effective practice
5. Provides coaching and expert support
6. Offers feedback and reflection
7. Is of sustained duration

The days of "sit-and-get" professional development are slowly being replaced with experiences that are job-embedded, meaningful for teachers, and focused on enhancing student learning. But there still is work to be done to ensure that all teachers are engaged in opportunities that maximize their professional learning. The fact that you are in a district that values the work of mathematics coaches is a step in the right direction!

In their seminal research, Joyce and Showers (2003) showed that including coaching as part of professional development experiences was critical to supporting teachers in meaningful professional learning. Joyce and Calhoun (2016) built upon this important work and found that a combination of four components was necessary for teachers to incorporate new learning into practice in meaningful ways: (a) understanding the rationale of a new practice; (b) seeing the practice in action; (c) planning for the new practice; and (d) participating in follow-up coaching. In fact, these researchers found that when (d) was neglected, just 5 percent to 10 percent of teachers actually implemented the learning long term. Conversely, when coaching was included, over 90 percent of teachers internalized the learning, and it became a part of their classroom practice. That is why *you*, the coach, are so important as part of an effective professional development program!

So we know from the discussion so far what effective teacher professional development should include, but what are your options when it comes to presenting professional development? Sometimes, you are responsible for presenting information, and you incorporate the previous suggestions as appropriate for the topic (active learning, opportunities to practice, follow-up coaching, etc.) But at other times, you are more of a facilitator of learning, and this can look very different. In many cases, this might be formalized (such as working with PLCs; see Chapter 13), but not always. There are many activities that fall into the category of professional development in which you might be a facilitator and not a presenter: lesson study, action research, book studies, co-teaching, analyzing student work (see Chapter 7), webinars, and mathematics studio (Gibbons, Lewis, & Batista, 2016), to name a few. In fact, many books written on professional development list coaching as a model of professional development. Of course, we believe coaching can be a very powerful form of professional development! Given the many professional development options, how do you decide which options are best for your teachers? Ask them! Teachers should be involved in making decisions about the professional development in which they will engage. Doing so honors their voices, builds trust, and solidifies commitment to the process.

Evaluating Professional Development

An essential part of professional development that is sometimes overlooked in the busy lives of coaches is the evaluation (Guskey, 2017). We know from adult learning theory (Knowles, 2015) that adults need to be involved not only in the planning of professional learning experiences but also in the evaluation of those experiences. Evaluating professional development can be done informally by simply asking teachers what worked and what didn't (e. g., exit slips, pluses and wishes). However, sometimes you need a more formal evaluation of the professional development, including data on teacher practice and student outcomes. Of course, this looks really different than exit slips. Many times, this could involve external evaluators examining a professional development program across one or more school years. So the range of evaluation techniques is varied and depends on the type of data that is needed.

There are many resources available to support you in learning about evaluating professional development (see Where to Learn More), but we will focus on Guskey's (2000) suggestion of the five areas that should be considered when evaluating professional development: (a) participants' reactions (ask the teachers!), (b) participants' learning, (c) organizational support and change, (d) participants' use of new knowledge and skills, and (e) student learning outcomes. Tool 12.5 includes these five areas and then provides additional support in terms of questions that can be asked, ideas for data collection, and suggestions for using the data for each of the five areas.

Tips for Presenting Professional Development

This section focuses on two of the seven characteristics of effective professional development (Darling-Hammond et al., 2017) listed earlier: (a) incorporates active learning and (b) supports collaboration. We know from Chapter 4 (Engaging Students) that engagement techniques are important in supporting high levels of learning for students—the same can be said for adults! Here, we list specific tips (strategies) you can use to actively engage participants and support collaboration in professional development (Aguilar, 2016; Garmston & Wellman, 2016).

Incorporates Active Learning

- *Line Up.* There are many ways to use this strategy, and you will think of many more! Ask participants to line up in alphabetical order, based on the answer to a question (i.e., What profession would you have been in if you have not chosen education?). An alternative is to ask participants to line up according to their birth month and day without talking. Once participants are lined up, you can have them count off to make groups of any size. This activity brings energy to the room and facilitates participants hearing the perspectives of others in the room.

- *Learning Partners.* This strategy also has many variations. To illustrate here, we will use Shape Partners. Create a handout with an icon for four shapes (see Tool 12.7). Ask participants to roam the room and find a partner for each shape. When you need participants in pairs, ask them to find their trapezoid partner. Tool 12.6 is another example of a learning-partner strategy.

Supports Collaboration

- *Say Something.* Participants collaborate with a partner. They each read a designated section of information, then each "say something" about what they just read. Continue with this process until the reading selection is completed. This works best with relatively short readings.
- *First Turn, Last Turn.* In groups of four, participants read a section of text and highlight key ideas for them. In turn, members share one of their key ideas but do not elaborate—they just read it. Next, each participant comments on the idea, but there can be *no cross talk*—so no dialogue at this point, just each participant sharing her or his thoughts about the idea. After all have shared, the person who began shares his or her thinking about the idea and the other ideas shared and also gets the "last turn." The pattern repeats until everyone has shared a key idea. This strategy is great for focusing on listening to others and getting all voices in the room.

Coaching Lessons From the Field

We have used a sequenced activity that includes a vignette to help teachers focus on both the Teaching Practices (TPs) and Mathematical Practices (MPs). The sequence includes (a) working through a mathematical problem or task, (b) analyzing student work with the vignette, and then (c) reflecting. We have successfully used this activity with both preservice and inservice teachers. We have found that as teachers engage in reading the vignette, analyzing student work, and reflecting on their own practice, they gain a more in-depth understanding about what those practices mean for them and for their students, thus impacting planning and instruction. Often, participants initially have a surface understanding of the Teaching Practices, but when given a focused activity in which they analyze and discuss, they become better able

to identify the effective and ineffective use of these practices and see them through the lens of their own practice. The most empowering part of the process is when they eventually write their own mini-vignette about an occurrence in their own practice/classroom and share it in discussions with their peers. Initially, looking at mini-vignettes we created together (such as the ones in Tool 12.7) is helpful and safe for the teachers. They are less intimidated and more willing to analyze that situation and then connect to their own practice. But once they are more comfortable with the sequence, it becomes more valuable for them to bring in their own mini-vignettes and connect directly to their own practice in a safe community environment.

The sequence is illustrated in the following visual:

Solve Task → Review Vignette of Students Solving Task → Identify MPs and MTPs → Analyze Student Work → Reflect and Connect to Own Practice

— Trena Wilkerson, Keith Kerschen, Ryann Shelton
Baylor University

 Go to **resources.corwin.com/mathematicscoaching** to download a sample vignette you can use in this sequence.

Where to Learn More

Books

Costa, A. L., & Wellman, B. M. (2016). *Adaptive Schools: A Sourcebook for Developing Collaborative Groups* (3rd ed.). Lanham, MA: Rowman & Littlefield.

> *This book, in its third edition, contains 35 specific facilitation moves for you to use in professional learning experiences. The moves will support you as you (a) focus and maintain attention, (b) manage energy, and (c) group participants. The authors also identify which moves in particular are helpful in dealing with problems that can arise during presentation and facilitation. Also included are 125 facilitation strategies to keep participants actively engaged during professional development. These moves and strategies are updated on the Thinking Collaborative website (see Online Resources).*

Haslam, M. B. (2010). *Teacher Professional Development Evaluation Guide.* Oxford, OH: National Staff Development Council (Learning Forward).

> *This resource contains information on the five questions to inform evaluation, design and data collection strategies, data quality, and data-reporting suggestions. Also included are resources for evaluating professional development and sample survey questions.*

Stiles, K. E., Mundry, S., Loucks-Horsley, S., Hewson, P. W., & Love, N. (2010). *Designing Professional Development for Teachers of Science and Mathematics* (3rd ed.). Thousand Oaks, CA: Corwin.

> *This valuable book, in its third edition, provides a comprehensive look at designing mathematics professional development. Chapters are included on the knowledge, beliefs, and contextual factors that influence professional development, as well as a design framework. Case studies of the framework in action tie it all together.*

Taylor, K., & Marienau, C. (2016). *Facilitating Learning With the Adult Brain in Mind: A Conceptual and Practical Guide.* San Francisco, CA: Jossey-Bass.

> *The book presents an overview of brain research and the implications for adult learners. The authors provide several "brain-aware" strategies to support meaningful learning experiences for adults.*

Yendol-Hoppey, D., & Dana, N. F. (2010). *Powerful Professional Development: Building Expertise Within the Four Walls of Your School.* Thousand Oaks, CA: Corwin.

> *This book describes how to use the expertise in your building to create powerful job-embedded professional learning experiences. The authors describe how to create a "theory of change" that will undergird your professional development efforts. Seven chapters focus on a variety of strategies for professional development that will meet a range of needs.*

Articles

Guskey, T. (2012). "The Rules of Evidence." *Journal of Staff Development* 33(4), 40–43. *Guskey presents five issues related to gathering evidence on the effectiveness of professional development.*

Learning Forward

www.learningforward.org

The website of Learning Forward, The Professional Learning Association has useful information and resources. While not specific to mathematics, their Standards for Professional Learning offer solid research-based practices to support your work. Other features include a Twitter feed, blog, publications, and information on their annual conference.

NCSM, Leadership in Mathematics Education

www.mathleadership.org

NCSM has done much work in the last few years to reach out and support mathematics coaches. Their mathematics leadership framework, Themes and Imperatives for Mathematics Education (TIME), contains four leadership principles that can guide your work as a coach. Their website has a special feature just for you called the Coaching Corner. This resource includes videos, resources for coaches, professional learning for coaches, and links to useful websites.

The Thinking Collaborative

www.thinkingcollaborative.com

This website contains over 160 strategies and moves you can use during professional learning experiences. Go to Resources, *then* Strategies. *The site is updated periodically.*

Coach's Toolkit

These tools are a menu from which you can select any that make sense for your setting/context.

Plan

12.1	Professional Development Overview Planning
12.2	Professional Development Planning Checklist
12.3	Professional Development Planning Template
12.4	Differentiating Professional Development

Evaluate

12.5	Evaluating Professional Development

Sample PD Activities

12.6	Sample Professional Development Activity: Grouping Strategy (P.I.C.S. Page)
12.7	Sample Professional Development Activity: Grouping Strategy (Shape Partners)
12.8	Sample Professional Development Activity: Mathematical Practices Mini-Vignettes
12.9	Sample Professional Development Activity: Mathematical Practices Playing Cards
12.10	Sample Professional Development Activity: Questioning and the Mathematical Practices
12.11	Sample Professional Development Activity: Effective Teaching Practices

 To download the coaching tools for Chapter 12 go to **resources.corwin.com/ mathematicscoaching**.

12.1 Professional Development Overview Planning

Instructions to the Coach: Use this tool to help you focus on the big ideas when planning professional development. Then, use *Tool 12.2 or 12.3 to plan the specifics.*

Mathematics Content	Teacher Knowledge	Resources
What mathematics content is the focus of the PD? What are students' needs relative to this content?	What pedagogical content knowledge (PCK) is the focus of the PD? What are the teachers' needs relative to this PCK?	What resources might you include to support the content and teacher knowledge goals (e.g., using children's literature, interactive SMART board, GeoGebra)?

Source: Previously published by Bay-Williams, J., McGatha, M., Kobett, B., and Wray, J. (2014). Mathematics Coaching: Resources and Tools for Coaches and Leaders, K–12. *New York, NY: Pearson Education, Inc.*

12.2 Professional Development Planning Checklist

☐ *Outcomes*. Explain what participants will know or be able to do as a result of the PD.

☐ *Make a connection* between the PD topic and system/school goals/initiatives.

☐ *Opener*. Relate the opener to your topic. Provide an opportunity for participants to briefly interact and "break the ice."

☐ *Model practices*. Include opportunities to model and have participants experience the practices you are addressing. Make connections to NCTM's Effective Teaching Practices, as appropriate.

☐ *Chunk your session to include these blocks:*

    • Prime Time 1—the first opportunity participants will have to learn new information

    • Processing Time—time for small groups/pairs to interact and process new material

    • Prime Time 2—a second opportunity for participants to learn new information (e.g., whole group debriefing/sharing of ideas)

☐ *Include small group interaction/processing/discussion*. Do this two to three times during the session.

☐ *Include examples* of classroom practices and practical ideas teachers can walk away with and use right away.

☐ *Closure/reflection*. Model effective closure reflection—use structures that could be used with students. Include an opportunity for participants to identify how they will use/act on the new information.

Source: Developed by Corrine Gorzo, Howard County Public Schools, MD (2009). Previously published by Bay-Williams, J., McGatha, M., Kobett, B., and Wray, J. (2014). Mathematics Coaching: Resources and Tools for Coaches and Leaders, K–12. *New York, NY: Pearson Education, Inc.*

12.3 Professional Development Planning Template

Professional development focus or topic:	PD date:

Intended audience:

Outcomes: (What will participants know and be able to do as a result of the training?)

-
-
-

1. *Opener* (Brief activity to jump-start thinking and/or interaction among the participants)	**2.** *Introduce new content* (How will participants learn the new material?)

3. *Process the learning* (How will participants process and internalize the new information? What interactive structures can be used?)

4. *Sharing and debriefing the content* (How will participants share their ideas/reactions to the new material?)

5. *Reflection and closure* (How will the learning be summarized and personalized?)	**6.** *Follow-up* (What types of follow-up and support will be provided?)

Source: Adapted from Maryland State Department of Education. (2008). Maryland Teacher Professional Development Planning Guide. Retrieved from http://mdk12.org/share/pdf/MarylandTeacherProfessionalDevelopmentPlanningGuide.pdf. Previously published by Bay-Williams, J., McGatha, M., Kobett, B., and Wray, J. (2014). Mathematics Coaching: Resources and Tools for Coaches and Leaders, K–12. New York, NY: Pearson Education, Inc.

12.4 Differentiating Professional Development

Instructions to the Coach: Brodesky, Fagan, Tobey, and Hirsch (2016) describe a model to **differentiate professional development** *(DPD Model). They provide these tools to assist coaches in the process (pp. 34–35). They suggest that 60 percent of PD activities be core and 40 percent be differentiated. See Where to Learn More for additional details.*

Guiding Questions for the DPD Model		
PD planning questions	**What content will be *core* for all participants?**	**What content will be *differentiated?***
What are the professional learning goals? What are participants' professional learning needs?	What is essential for everyone to learn? For which areas do participants have a lot of consistency in their professional learning needs?	In what ways do the goals vary for different groups of teachers (by role, grade level, etc.)? For which areas do participants have a lot of variation in their professional learning needs? What is the distribution of needs?
What activities will you use to address the learning goals and participants' needs?	How important is it for all teachers to experience this activity for building knowledge and/or providing a shared experience?	What are ways to differentiate the activity to address teachers' varied needs? What choices might you offer?

Differentiated PD Planning Tool			
What are the professional learning targets?			
What are participants' learning needs?			
Time	**What are the topics and activities?** Fill out as you would for any agenda.Then, star (*) topics/ activities for which participants have particularly varied needs.	**What might be core (C) or differentiated (D)?**	**What will you differentiate? How?** Look over the topics/ activities that you marked *D*. Which of these are high priorities to differentiate? Choose a few and brainstorm ways to differentiate by using a choice point or other methods. Write down ideas below.

Source: Brodesky, A. R., Fagan, E. R., Tobey, C. R., & Hirsch, L. (2016). "Moving Beyond One-Size-Fits-All PD: A Model for Differentiating Professional Learning for Teachers." Journal of Mathematics Education Leadership, 17 (1), 20–37.

 Available for download at **resources.corwin.com/mathematicscoaching.** Copyright © 2018 by Corwin.

12.5 Evaluating Professional Development

Evaluation Level	What Questions Are Addressed?	How Will Information Be Gathered?	What Is Measured or Assessed?	How Will Information Be Used?
1. Participants' reactions	• Did they like it? • Was their time well spent? • Did the material make sense? • Will it be useful? • Was the leader knowledgeable and helpful? • Were the refreshments fresh and tasty? • Was the room the right temperature? • Were the chairs comfortable?	• Questionnaires administered at the end of the session	• Initial satisfaction with the experience	• To improve program design and delivery
2. Participants' learning	• Did participants acquire the intended knowledge and skills?	• Paper-and-pencil instruments • Simulations • Demonstrations • Participant reflections (oral and/or written) • Participant portfolios	• New knowledge and skills of participants	• To improve program content, format, and organization
3. Organizational support and change	• What was the impact on the organization? • Did it affect organizational climate and procedures? • Was the support public and overt? • Were problems addressed quickly and efficiently? • Were sufficient resources made available? • Were successes recognized and shared?	• District and school records • Minutes from follow-up meetings • Questionnaires • Structured interviews with participants and district or school administrators • Participant portfolios	• The organization's advocacy, support, accommodation, facilitation, and recognition	• To document and improve organizational support • To inform future change efforts

Available for download at **resources.corwin.com/mathematicscoaching**. Copyright © 2018 by Corwin.

260   Navigating a Successful Journey

Evaluation Level	What Questions Are Addressed?	How Will Information Be Gathered?	What Is Measured or Assessed?	How Will Information Be Used?
4. Participants' use of new knowledge and skills	• Did participants effectively apply the new knowledge and skills?	• Questionnaires • Structured interviews with participants and their supervisors • Participant reflections (oral and/or written) • Participant portfolios or products • Direct observations • Video or audio tapes	• Degree and quality of implementation	• To document and improve the implementation of program content
5. Student learning outcomes	• What was the impact on students? • Did it affect student performance or achievement? • Are students more confident as learners? • Is student attendance improving? • Are dropouts decreasing?	• Student data • School records • Questionnaires • Structured interviews with students, parents, teachers, and/or administrators • Participant portfolios	• Cognitive (performance & achievement) • Affective (attitudes & dispositions) • Psychomotor (skills & behaviors)	• To focus and improve all aspects of program design, implementation, and follow-up • To demonstrate the overall impact of professional development

Source: Guskey, T. R. (2000). *Evaluating Professional Development.* Thousand Oaks, CA: Corwin.

12.6 Sample Professional Development Activity: Grouping Strategy (P.I.C.S. Page)

*Instructions to the Coach: All materials for this activity can be downloaded at **resources.corwin. com/mathematicscoaching**.*

Professional development focus or topic: Getting to know the group while introducing the idea of making connections and building understanding of mathematics
Intended audience: Whole group of teachers (any grade)
Outcomes: Teachers will be able to …   • Meet colleagues participating in the professional development   • Discuss connections among representations for a mathematical situation   • Consider how a similar version of the activity might be used with their own students
Preparation: • Use the P.I.C.S. Page (Tool 3.4).   • Select enough topics so that there are four teachers per topic (e.g., if you have a group of 24 teachers, you will need six topics).   • Write each topic phrase (e.g., subtracting whole numbers, multiplying fractions, multiplying binomials, etc.) at the top of the page and make three copies (so you will have four pages for each topic). For each, highlight a different section (P, I, C, or S). See the miniature pages that follow this table for reference.

Description of activity:
 • Stack a "mixed" set of P.I.C.S. Pages on each table.
 • Invite participants to select one page and complete the highlighted section *only*.
 • On your cue (once participants seem to be finished), explain that they are to find the rest of their P.I.C.S. Topic Group.
 • Once they have found their group, ask them to do the following:
    ◦ Round 1: Share their name and their section of the P.I.C.S. Page. Encourage listeners to record what they hear and add it to their own page.
    ◦ Round 2: Describe one connection they see between their original section and someone else's section.
    ◦ Round 3: How might they use this page with their own students?
 • Debrief by asking participants to share any insights they gained (mathematically or otherwise) from doing this activity.
 	
 Note: You can use this activity to form groups for the rest of the session, or you can have participants return to their original table group.

P.I.C.S. Page Categories	
Procedure	Illustration
Concept	Situation

12.7 Sample Professional Development Activity: Grouping Strategy (Shape Partners)

Professional development focus or topic: Strategy for grouping participants in order to hear multiple perspectives and to include physical movement.
Intended audience: Whole group of teachers (any grade)
Outcomes: Teachers will be able to …   • Meet colleagues participating in the professional development   • Understand perspectives of other participants
Preparation: • Copy the Shape Partners handout provided on the next page, one for each participant.
Description of activity: • Give a Shape Partners handout to each participant.   • Explain that they will roam around the room and identify four people (not currently sitting at their table) to be their shape partners. For example, find a partner that has a shape available. Each participant writes the other's name on that shape. When they have different names for each shape, they can return to their seat.   • Sometimes, it will work out that not everyone gets a name for all four shapes. Explain that this is fine. When you ask participants to find their trapezoid shape, for example, if they do not have a name or their partner is absent, they can come to the front of the room and you can facilitate finding partners. If everyone is partnered, ask them to join a group to make a trio.   • This strategy can be used for stand-up conversations, or partners can find a place to sit with their new partner for an extended activity.   *Note*: You can use this activity to form groups for the rest of the session, or you can have participants return to their original table group.

NOTES

_____
_____
_____
_____
_____
_____
_____
_____
_____
_____
_____
_____
_____
_____
_____
_____
_____
_____
_____
_____

Shape Partners For Tool 12.7

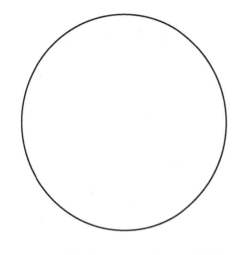

_____

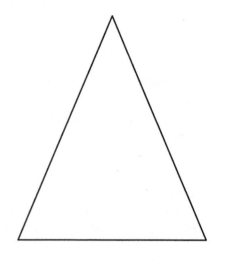

_____

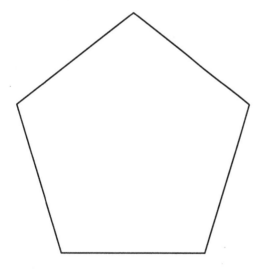

_____

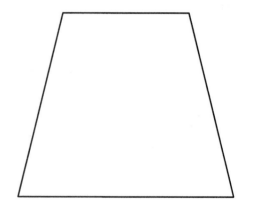

_____

12.8 Sample Professional Development Activity: Mathematical Practices Mini-Vignettes

Professional development focus or topic: Deepening understanding of the Mathematical Practices
Intended audience: Small group or large group of teachers (any grade)
Outcomes: Teachers will be able to …   • Connect student actions with the Mathematical Practices   • Describe the essence of each of the Mathematical Practices   • Describe distinctions and connections among the Mathematical Practices

Preparation:
 • Copy the mini-vignettes on bright paper, laminate (optional), and cut into cards.
   ◦ Elementary Mini-Vignettes (on next page, for example, and available for download)
   ◦ Middle School Mini-Vignettes (available for download)
   ◦ High School Mini-Vignettes (available for download)
 • Copy the placemat (one for every group of teachers) and laminate (optional).
 • Download the Mathematical Practices & Student Look Fors Bookmark for each teacher (optional).

Description of activity:
 • Distribute the placemats and bookmarks to teachers. If needed, provide time for teachers to read each of the Mathematical Practices.
 • Place teachers in small groups (3–4).
 • Instruct groups to have each teacher draw one of the mini-vignettes, read it silently, and decide which Mathematical Practice he or she thinks it *best* fits.
 • After everyone is ready, instruct teachers to go one by one, sharing their vignette and placing it on the placemat where that teacher thinks it should go. Allow time for other teachers to question if it could go in another spot on the placemat. Repeat until each mini-vignette has been placed on the placemat. (You can refer to the vignettes by the student's name in the vignette to reinforce that the Mathematical Practices are *student* practices.)
 • If there are several groups of teachers, have them compare their results, sharing where they placed each mini-vignette.
 • Summarize by highlighting what has emerged from the conversation—likely that student actions (even small ones) can indicate more than one practice; that there are distinct meanings to each of the Mathematical Practices; and that the first step in being able to nurture the development of these practices is thinking about what they actually look like in the context of a math lesson.
 	
 Note: The vignettes were designed for one particular Mathematical Practice, so teachers should be able to place one card in each of the eight practices on the placemat. But they may do this in more than one way because the vignette intended for one practice has hints of other practices as well. Here are our intended matches:
 	
        SMP#1 – D   SMP#2 – H   SMP#3 – A   SMP#4 – E
 	
        SMP#5 – B   SMP#6 – F   SMP#7 – C   SMP#8 – G

Mini-Vignettes (Elementary) For Tool 12.8

D. Anna is trying to find the area of an unusually shaped garden. She thinks about a simpler problem of a rectangular garden. She partitions the garden into familiar shapes to solve the task. As she works, she monitors and evaluates her progress and adapts her strategy when it doesn't seem to be working.

H. Christopher is working on addition strategies. He looks at 8 + 7 and decides to use the context of his toy cars to think about the problem. He recognizes that 8 cars equals 5 cars and 3 more and 7 cars equals 5 cars and 2 more. He pictures the cars lined up in fives and solves the problem by adding 5 + 5 + 5.

A. Noah is studying the attributes of various triangles. He says, "A scalene triangle cannot have a right angle." Nick says, "Yes, it can, as long as the sides are different lengths." Amy says, "It is an equilateral triangle that cannot have a right angle." Maria says, "An obtuse triangle can't have a right angle either."

E. Rachel listened to a story about a family of four who wanted to grow their family to 10. She decides to represent the story using this equation: 4 + c = 10 and draws a number line to show the situation:

B. To solve $\frac{3}{4} + \frac{3}{8}$, Mindy decides to use a number line. She starts at $\frac{3}{4}$, which is also $\frac{6}{8}$, jumps up $\frac{2}{8}$ to get to 1, and then one more eighth to get to $1\frac{1}{8}$.

F. Lin and Ben are working on describing what makes a square a square. Lin says, "All the sides are the same." Ben says, "But that is true for a rhombus, too." Lin pauses and changes his description, saying, "It is a quadrilateral with four equal sides and four right angles."

C. In working on expressions such as 6 x 3 x 5, Zöe realizes that she gets the same answer if she multiplies 6 x 5 x 3, which is easier to do in her head. She realizes that this will always work because each of these factors could be the measures of the sides of a box, which can be in any position.

G. José is using a hundreds chart to count by 10, starting with various numbers. When starting at 28, he recognizes that all the numbers end in 8 (38, 48, 58, etc.). When the teacher asks him to add 9 to 28, he notices that the numbers are one less than counting by tens (37, 47, 57, etc.), and they all end in 7.

Mathematical Practices Placemat For Tool 12.8

1. Make sense of problems and persevere in solving them.
Mathematically proficient students start by explaining to themselves the meaning of a problem and looking for entry points to its solution.

2. Reason abstractly and quantitatively.
Mathematically proficient students make sense of quantities and their relationships in problem situations.

3. Construct viable arguments and critique the reasoning of others.
Mathematically proficient students understand and use stated assumptions, definitions, and previously established results in constructing arguments.

4. Model with mathematics.
Mathematically proficient students can apply the mathematics they know to solve problems arising in everyday life, society, and the workplace.

5. Use appropriate tools strategically.
Mathematically proficient students consider the available tools when solving a mathematical problem.

6. Attend to precision.
Mathematically proficient students try to communicate precisely to others.

7. Look for and make use of structure.
Mathematically proficient students look closely to discern a pattern or structure.

8. Look for and express regularity in repeated reasoning.
Mathematically proficient students notice if calculations are repeated and look both for general methods and for shortcuts.

12.9 Sample Professional Development Activity: Mathematical Practices Playing Cards

Instructions to the Coach: All materials can also be downloaded at **resources.corwin.com/ mathematicscoaching**.

Professional development focus or topic: Deepening understanding of the Mathematical Practices
Intended audience: Small group or large group of teachers (any grade)
Outcomes: Teachers will be able to … • Connect student actions with the Mathematical Practices • Describe the essence of each of the Mathematical Practices • Describe distinctions and connections among the Mathematical Practices
Preparation: • Copy the MP Playing Cards on the next two pages bright paper and cut into cards. • Download the Mathematical Practices & Student Look Fors Bookmark for each teacher (optional).
Description of activity: • Distribute one playing card and a bookmark to each teacher. If needed, provide time for teachers to read the look fors for each of the Mathematical Practices on the bookmark. • Instruct teachers to create their own mini-vignette that illustrates the MP on their playing card. When scaffolding this activity for teachers, you may want to do the mini-vignette activity first (Tool 12.7), so teachers have an idea what you mean by *mini-vignette*. • After everyone is ready, facilitate teachers forming groups of eight. This can be accomplished several ways: you can seat teachers in groups of eight before you hand out the MP playing cards; you can copy the playing cards on different-colored paper and ask teachers to find seven others who have the same color paper; or you can put colored dots or stickers of some kind on the back of the cards before you hand them out and ask teachers to find seven people who have a red dot, for example, on the back of their playing card. The group of eight should have teachers with MP playing cards 1 through 8, so all MPs are represented. • Optional: Depending on the size of the group, you could add a step prior to the small-group sharing of having expert groups (e.g., all the MP 1 teachers) form a group and share their vignettes to allow revising if necessary. Then, proceed to small-group sharing. • Once in their groups, ask teachers to read their vignette and let others in the group decide which MP it illustrates. Allow time for discussion. • Summarize by highlighting what has emerged from the conversation—likely that student actions (even small ones) can indicate more than one practice; that there are distinct meanings to each of the Mathematical Practices; and that the first step in being able to nurture the development of these practices is thinking about what they actually look like in the context of a math lesson.

Mathematical Practices Playing Cards For Tool 12.9

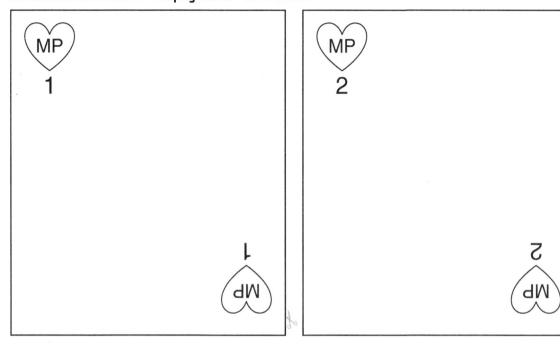

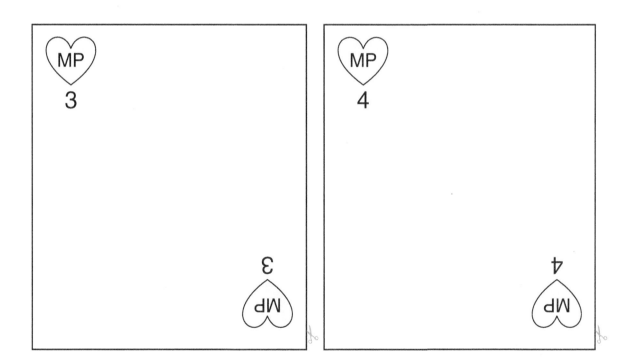

Mathematical Practices Playing Cards For Tool 12.9

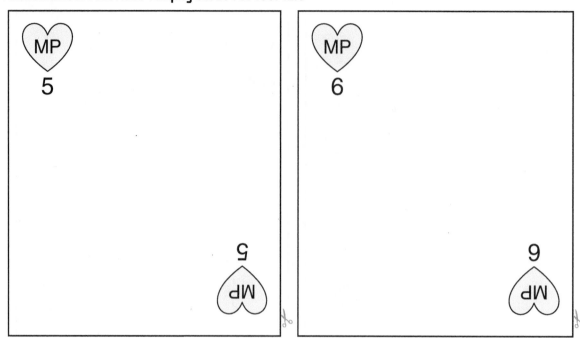

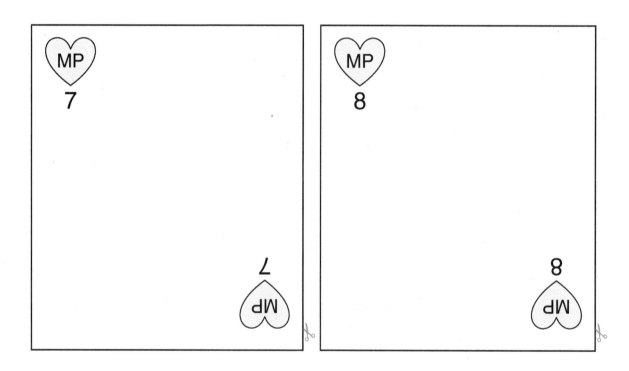

12.10 Sample Professional Development Activity: Questioning and the Mathematical Practices

Instructions to the Coach: All materials for this activity can be downloaded at **resources.corwin.com/mathematicscoaching**.

Professional development focus or topic: Deepening understanding of asking questions to provide opportunities for students to demonstrate the Mathematical Practices
Intended audience: Small group or large group of teachers (any grade)
Outcomes: Teachers will be able to … • Connect questions for students with the Mathematical Practices
Preparation: • Copy the sample questions on colored paper and laminate (optional). Cut apart and put each set into an envelope. You will need one set of questions for each group. • Download the Mathematical Practices & Student Look Fors Bookmark for each teacher (optional). • Copy and laminate (optional) one Mathematical Practices placemat for each group.
Description of activity: • Distribute one Mathematical Practices placemat and a bookmark to each teacher. If needed, provide time for teachers to read the look fors for each of the Mathematical Practices on the bookmark. • Distribute one set of sample questions to each group. • Instruct each participant to read a question to the group and discuss which Mathematical Practice students might have an opportunity to demonstrate if asked the question. Instruct participants to place the question in the appropriate place on the placemat. Continue until all questions have been placed on the placement. • Optional: You can have participants create their own questions to add to the placemat. • Summarize by highlighting what has emerged from the conversation. Share the answer key, which is merely one possible set of answers.

NOTES

_____

_____

_____

_____

_____

_____

_____

_____

_____

_____

_____

_____

_____

_____

_____

Sample Questions For Tool 12.10

How is this task similar to a previous task you have completed?	How might a number line help you think about the problem?
What strategies might help you to solve this problem?	What manipulative or picture might you use to solve the problem?
What helped you be successful in solving the problem?	What other resources might help you with this problem?
What expression or equation represents this data/situation?	When will this strategy work?
Are these expressions equivalent? How do you know?	Which is the better unit of measure for this task?
What do the variables/numbers/answer mean related to the context?	What labels might be useful for this problem?
Why did you use [a graph] to solve it?	When can you regroup numbers and maintain equivalence?
How did you get [that equation]?	How might you use **break-apart** to solve this problem?
What do the rest of you think about Anna's strategy?	What is true about all of these triangles?
How does your model/equation connect to the situation?	What patterns do you notice across these problems?
Where can you find [the rate] in this situation? The table? The equation?	How are these problems the same? Different?
Are these two equations equivalent? Which (if any) is more efficient?	How might this problem help you solve another problem?

Mathematical Practices Placemat	
1. Make sense of problems and persevere in solving them.	
2. Reason abstractly and quantitatively.	
3. Construct viable arguments and critique the reasoning of others.	
4. Model with mathematics.	
5. Use appropriate tools strategically.	
6. Attend to precision.	
7. Look for and make use of structure.	
8. Look for and express regularity in repeated reasoning.	

Answer Key For Tool 12.10 [Note: This is one possible set of answers.]

1. Make sense of problems and persevere in solving them.	How is this task similar to a previous task you have completed? What strategies might help you to solve this problem? What helped you be successful in solving the problem?
2. Reason abstractly and quantitatively.	What expression or equation represents this data/situation? Are these expressions equivalent? How do you know? What do the variables/numbers/answer mean related to the context?
3. Construct viable arguments and critique the reasoning of others.	Why did you use [a graph] to solve it? How did you get [that equation]? What do the rest of you think about Anna's strategy?
4. Model with mathematics.	How does your model (equation) connect to the situation? Where can you find [the rate] in this situation? The table? The equation? Are these two equations equivalent? Which (if any) is more efficient?
5. Use appropriate tools strategically.	How might a number line help you think about the problem? What manipulative or picture might you use to solve the problem? What other resources might help you with this problem?
6. Attend to precision.	When will this strategy work? Which is the better unit of measure for this task? What labels might be useful for this problem?
7. Look for and make use of structure.	When can you regroup numbers and maintain equivalence? How might you use break-apart to solve this problem? What is true about all of these triangles?
8. Look for and express regularity in repeated reasoning.	What patterns do you notice across these problems? How are these problems the same? Different? How might this problem help you solve another problem?

12.11 Sample Professional Development Activity: Effective Teaching Practices

Instructions to the Coach: All materials can also be downloaded at **resources.corwin.com/ mathematicscoaching**.

Professional development focus or topic: Deepening understanding of the NCTM's *Effective Teaching Practices*

Intended audience: Small group or large group of teachers (any grade)

Outcomes: Teachers will be able to …
- Connect teacher actions with the Teaching Practices
- Describe the essence of each of the Teaching Practices
- Describe distinctions and connections among the Teaching Practices

Preparation:
- Download the Teaching Practices & *Shifts in Classroom Practice* Look Fors Bookmark for each teacher.
- Prepare *Shifts* Placemats by using Tools 2.1 and 2.10. Enlarge each *Shift* so there is one at the top of each page (horizontal format). This can also be done on 11x14 paper so there is more space or even as large as chart paper. You should have one placemat for each *Shift*.

Description of activity:
- Distribute a bookmark to each teacher. If needed, provide time for teachers to read the look fors for each of the Teaching Practices on the bookmark.
- Give a *Shifts* Placemat to each small group of teachers.
- Instruct teachers to generate examples of teacher actions (what a teacher might say or do during a lesson) for their *Shift* and write each on a sticky note and place it where it might fall on the continuum.
- Encourage teachers to generate teacher moves for the entire continuum, not just the right end.
- After each group has completed its list of teacher actions, engage teachers in a gallery walk to view the other *Shifts* continua. You can require each table group to add one more example for each *Shift*, if time allows. Alternatively, you could ask each group to share their teacher actions with the whole group instead of the gallery walk.
- When the gallery walk is concluded, debrief about what teachers noticed. You could ask them to do a self-assessment of where they see their teaching on specific or all *Shifts* (see Tool 2.1).
- Next, engage teachers in a dialogue about what Mathematical Practices could be supported by the teacher actions. Share a teacher action to illustrate for the group. For example, a teacher might say, "You switched strategies—what made you decide to try something different?" This would be located toward the right end of the continuum for *Shift 7*. Now the question becomes this: If a teacher asked this question, which Mathematical Practice might students have an opportunity to demonstrate? This teacher action seems to align with Mathematical Practice 1 (Make sense of problems and persevere in solving them). Have table groups return to the teacher actions they created for their *Shift*. On each sticky note, indicate which Mathematical Practice could be demonstrated by students.
- Summarize by highlighting what has emerged from the conversation. This activity is powerful because teachers begin to see that there are very few teacher actions on the left end of the continua that connect to Mathematical Practices! It is an important reminder of why teaching needs to move toward the right end of the continua.

Chapter 13

Facilitating Professional Learning Communities

Dear Coach,

Here is support for you as you work to increase your effectiveness in facilitating PLCs.

In the **COACH'S DIGEST** ...

Overview: A brief review of facilitating PLCs.

Tips: Strategies for effective facilitation of PLCs.

Coaching Lessons From the Field: Mathematics leaders share how they used a tool or idea from this chapter.

Where to Learn More: Articles, books, and online resources for you on where to learn more about facilitating PLCs!

In the **COACH'S TOOLKIT** ...

Five tools focused on facilitating PLCs.

Coach's Digest

In the Coach's Digest, we begin with an overview of facilitating professional learning communities to support you in one of the many roles in which you engage. As you read the Overview, the following questions might help you consider it in terms of your role as a coach:

- How do I know if I need to be a presenter or a facilitator? What is the difference?
- What meeting structures might I implement to make PLC meetings as efficient and effective as possible?

Overview of Facilitating Professional Learning Communities

There is a multitude of material written about **professional learning communities (PLCs)**. With all that material also comes a bit of confusion as authors use various names to describe slightly different nuances about groups of teachers working and learning together. The most widely used descriptor is *professional learning communities*, but there are also *communities of practice, professional communities learning, professional learning teams*, and *communities of learners*, to name just a few. The focus of this chapter is not on unpacking these distinctions; we will use *PLC*, as we have throughout this book, to mean a group of educators coming together to learn and grow in order to enhance performance. As mentioned in Chapter 12, you wear many hats as a coach, and providing professional development is likely one of your responsibilities. Many times, a large part of that professional development is focused on facilitating PLCs. So this chapter will focus on *how* to facilitate PLCs (or whatever you call them!).

Let's begin with making explicit the distinction between *presenter* and *facilitator* that we alluded to in Chapter 12. Presenting is teaching. A presenter supports the development of knowledge and skills that teachers can apply in their practice. For example, if you were presenting a professional development on the Mathematical Practices (perhaps using the PD activities included in Chapter 12!), your goal would be supporting teachers in a deeper understanding of the Mathematical Practices. You would be in front of the group, leading them in activities to develop new understandings—that is, presenting. Sometimes, the strategies and moves that a presenter uses are similar to, or even identical to, the strategies and moves of a facilitator. This can lead to confusion between the two roles. A facilitator creates an environment in which teachers feel safe to take risks and collaborate. The job of the facilitator is to direct the processes used in a meeting and to support trusting relationships in order to meet group goals (Garmston & Wellman, 2016; Killion, 2013). *Facilitate* means to "make easier," so a facilitator makes meetings (PLCs, faculty meetings, content-focused meetings, etc.) easier for the participants. Unfortunately, we have all been in meetings that we felt were a waste of time. Frequently, this was the result of poor facilitation. Learning to become a proficient facilitator is important! In fact, Garmston and Zimmerman (2013a, p. 11) suggest that "because time is the educator's most valuable asset, a moral imperative is that all leaders pay attention to and guarantee excellent facilitation and appropriate interventions." So if facilitation is that important, what do you need to be paying attention to?

There are entire books focused on group facilitation (e.g., Aguilar, 2016; Garmston & Zimmerman, 2013b; Killerman & Bolger, 2016). To support you in focusing on one specific aspect of facilitation that you might use with PLCs, we will concentrate on "developing meeting standards" from the *Adaptive Schools Sourcebook* (Garmston & Wellman, 2016). Any time you bring a group of diverse people together in a PLC, there is the potential for chaos! As a facilitator, you can implement structures to honor the diversity in the group and to keep the meeting as efficient and effective as possible. Developing meeting standards is one type of structure you can use; it is characterized by the five standards described in Figure 13.1. Then, in the next section we describe facilitation tips (strategies) for each meeting standard.

Figure 13.1   Develop Meeting Standards

One process at a time	Group processes are strategies to support the group in generating ideas, clarifying, analyzing, and problem-solving. With the variety of personalities that are present in any meeting, it is natural for group members to engage in this work with different approaches, which can potentially waste time as group members work in different directions. The facilitator can save the group time by helping them to focus on just one process at a time.
One topic at a time	You have probably experienced a meeting in which multiple topics were being discussed at the same time and can remember the confusion it caused. These types of discussions happen because one issue is connected with many others and it is easy to get off track. This can cause the group to lose focus and momentum. Supporting the group to focus on one topic at a time can reduce confusion and frustration, as well as make more efficient use of time.
Balance participation	In Chapter 12, we discussed the importance of actively engaging adults when presenting professional development. That same principle applies in meetings. You want to balance participation so everyone in the group is actively engaged, not just one or two people.
Understand and agree on roles	There can be a variety of roles in any group setting (e.g., facilitator, recorder, group member). These roles might change from time to time, and it is important for everyone to understand and agree on the roles to support effective and efficient meetings.
Engage cognitive conflict	Cognitive conflict is a critical aspect of effective groups. **Cognitive conflict** occurs with disagreements about important differences of opinion, and **affective conflict** occurs with disagreements over personalized anger or resentment. Cognitive conflict supports groups in making better decisions and produces increased commitment to the group. As a facilitator, you want to encourage cognitive conflict and discourage affective conflict.

Source: *Adapted from Garmston, R. J., and Wellman, B. M. (2016). The Adaptive School: A Sourcebook for Developing Collaborative Groups (3rd ed.). Lanham, MA: Rowman & Littlefield.*

Developing meeting standards is just one example of a structure you can use as a facilitator of PLCs to support the group in meeting its goals. As the PLC learns from its experiences, teachers will become more skillful in each of the meeting standards. It is important to occasionally lead the PLC in self-assessing its use of the meeting standards to monitor progress and identify areas of needed growth (Garmston & Wellman, 2016; see Tool 13.1).

Tips for Facilitating Professional Learning Communities

Facilitation tips (strategies) for each of the meeting standards are presented in Figure 13.2. There are many possibilities for each meeting standard; we provide just a few examples.

Figure 13.2 Facilitation Tips for Meeting Standards

Meeting Standard	Facilitation Tips
One process at a time	*Process as Given, Process as Understood (PAG/PAU).* The facilitator explains the processes (steps) for the activity the PLC will engage in, including what to do and what not to do. Then, the facilitator checks for understanding of the processes from the group (e.g., What are we doing first? What are the ground rules? How much time will we spend on this?). To see a brief video of this strategy being used with a group in a brainstorming activity, go to www.thinkingcollaborative.com/asvideo-6-3.
One topic at a time	*Relevancy challenge.* The facilitator or a group member can use this strategy to help the group stay on the proposed topic by simply asking, "What are the connections between your comment and our topic?" *Parking Lot.* When a group member proposes a comment that is not related to the topic, you can record the topic on a chart posted in the room and designated as the "parking lot." Clarify that the group can return to the topic at a later time. To see a brief video of this strategy being used with a group in a brainstorming activity, go to www.thinkingcollaborative.com/ASvideo-8-5.
Balance participation	*Appreciative Inquiry Interviews.* These interviews are a great way to involve all participants, and they have the added benefit of focusing on positive aspects of teachers' practice (see Tool 13.2 for an example of an **Appreciative Inquiry interview**; read the coaching vignette for how one coach used these interviews).
Understand and agree on roles	*Four Corners.* To support teachers in understanding the various roles of the PLC (e.g., facilitator, group member, recorder), write each role on chart paper and post in a corner of the room. Teachers self-assign to a group and write on the chart paper the "job description" of that role. Ask groups to go to each chart and add to or write clarifying questions about the descriptions. End with a discussion and agreement upon each role and its job.
Engage **cognitive conflict**	*Assumptions Wall.* In small groups, have teachers write their assumptions about a topic and then select one that most influences their behavior. Each teacher posts the assumption where the group can see. Teachers engage in a structured dialogue about the assumptions. This strategy can surface cognitive conflict and support the group in making better decisions. To see detailed directions for this strategy, go to www.thinkingcollaborative.com/wp-content/uploads/2016/10/Assumptions-Wall.pdf.

Source: Adapted from Garmston & Wellman (2016).

Coaching Lessons From the Field

I noticed that the teachers in one of the PLCs I facilitate focused on the most challenging aspects of their day. Five wonderful things could happen, but even if only one small negative thing happened, they couldn't let it go. It seemed these teachers could easily fall into an abyss of negativity and struggle to climb out. I truly valued how hard they were working, yet they were so hard on themselves! I decided to use a strengths-based approach called *Appreciative Inquiry* (Cooperrider & Whitney, 2005) to frame our discussions. While I wanted them to be able to face their challenges, I loved that Appreciative Inquiry (AI) focused on unpacking and understanding "peak experiences." I wanted to be able to help them understand that solving their challenges resided in knowing, understanding, and leveraging what they were doing well.

I decided to focus on mathematical discourse. I began by asking them to focus on a peak experience they had teaching mathematics when the students were highly engaged in discourse. I paired the teachers and asked them to interview each other. I encouraged them to unpack explicit details about that experience, including how they felt, the energy in the room, what they were doing, what students were doing, and what they valued about their role in this experience. From there, the pair joined another pair to share stories. The key was that each partner had to share his or her partner's story with the other pair! I also instructed the pairs to notice and record themes that were occurring in the stories. I stood in complete amazement as I watched the teachers share their successful stories with enthusiasm, laughter, and even wonder. High-cognitive tasks, real-world tasks, opportunities for students to share ideas, and mixed-ability grouping were common themes in the teachers' success stories. I posted all these wonderful themes and asked the participants, "What would your dream classroom look like if you could facilitate this kind of discourse all the time?" as well as, "Using the details of your success, how could you design your classroom to engage your learners in mathematical discourse?"

At once, the teachers began planning teaching experiments to engage their students in rich discourse. In some cases, teachers needed to learn more information, others decided to establish a new routine, and another teacher decided to strategically build and plan more opportunities for her students to talk throughout the lesson. At this time, I was able to introduce new ideas about facilitating classroom discourse that matched their strengths, needs, and motivation. Teachers are still talking about how energized they were by this experience!

Where to Learn More

Books

Aguilar, E. (2016). *The Art of Coaching Teams*. San Francisco, CA: Jossey-Bass.

> *This book is a how-to guide for building effective teams. The author includes many tools and community-building activities to use when facilitating groups.*

Garmston, R. J., & Wellman, B. M. (2016). *The Adaptive School: A Sourcebook for Developing Collaborative Groups* (3rd ed.). Lanham, MD: Rowman & Littlefield.

> *In its third edition, this outstanding book lays the groundwork for everything you need to know about facilitating and developing collaborative groups. Topics such as collaboration, successful meetings, dealing with conflict, and resources for inquiry will support you in becoming an outstanding facilitator.*

Garmston, R. J., & Zimmerman, D. P. (2013). *Lemons to Lemonade: Resolving Problems in Meetings, Workshops, and PLCs*. Thousand Oaks, CA: Corwin.

> *The authors provide a complete guide to dealing with challenging issues in group settings. They offer a variety of facilitation strategies and moves to support you in dealing with issues such as deciding when to intervene, managing common challenges, and strategies for advanced facilitation.*

Jolly, A. (2008). *Team to Teach: A Facilitator's Guide to Professional Learning Teams*. Oxford, OH: National Staff Development Council (Learning Forward).

> *This book offers over 97 tools to support you in facilitating PLCs. Available at* https://learningforward.org/docs/default-source/docs/teamtoteach-tools.pdf.

Killion, J. (2013). *School-Based Professional Learning for Implementing the Common Core, Unit 2, Facilitating Learning Teams*. Oxford, OH: National Staff Development Council (Learning Forward).

> *This resource is a comprehensive set of materials focused on professional development around implementing content standards. However, the second unit is focused specifically on facilitation and offers many tools and practical suggestions to support you as you facilitate PLCs. Available at https://learningforward.org/publications/implementing-common-core/professional-learning-units.*

Articles

von Frank, V. (2013). "Finding Your Voice in Facilitating Productive Conversations." *The Leading Teacher*, 8(4), 1, 4–5.

> *This quick read offers practical suggestions for facilitating productive discussions. The author includes sentence starts to use in difficult conversations.*

Online Resources

Facilitation Tools and Techniques at Learning for Sustainability

www.learningforsustainability.net/facilitation

> *Learning for Sustainability's website houses a variety of resources related to facilitation. While not specifically written for the education community, the resources are useful.*

Learning Forward

www.learningforward.org

> *The website of Learning Forward, The Professional Learning Association, has useful information and resources. If you go to* Popular Searches *and click on* Facilitation, *you will get numerous resources to support you as you facilitate PLCs.*

National School Reform Faculty

www.nsrfharmony.org

> *This website provides over 200 free protocols and activities to use when facilitating meetings in categories such as learning communities, equity, inquiry, team- and trust-building, dilemmas, observations, and brainstorming.*

The Thinking Collaborative

www.thinkingcollaborative.com

> *This webpage contains over 160 strategies and moves you can use during professional learning experiences. Go to* Resources, then Strategies. *It is periodically updated.*

NOTES

Coach's Toolkit

These tools are a menu from which you can select any that make sense for your setting/context.

Self-Assess

13.1   Meeting Standards Self-Assessment and Reflection

13.2   Facilitator Proficiency Scale

Sample Facilitation Activities

13.3   Sample Facilitation Activity: Appreciative Inquiry Protocol

13.4   Sample Facilitation Activity: Paraphrasing

13.5   Sample Facilitation Activity: Inclusion Strategies

13.6   Sample Facilitation Activity: Task Talk Protocol (Kobett et al., 2015)

 To download the coaching tools for Chapter 13 go to **resources.corwin.com/mathematicscoaching.**

13.1 Meeting Standards Self-Assessment and Reflection

Instructions to the Coach: Use this tool to help you self-assess your implementation of the meeting standards (Garmston & Wellman, 2016). When the meeting standards become a normal part of the PLC's activity, you can ask PLC members to do a self-assessment for the group.

Meeting Standard	Self-Assessment
1. One process at a time	Never ●————————————————● Always Sometimes
Evidence and next steps:	
2. One topic at a time	Never ●————————————————● Always Sometimes
Evidence and next steps:	
3. Balance participation	Never ●————————————————● Always Sometimes
Evidence and next steps:	
4. Understand and agree on roles	Never ●————————————————● Always Sometimes
Evidence and next steps:	
5. Engage cognitive conflict	Never ●————————————————● Always Sometimes
Evidence and next steps:	

13.2 Facilitator Proficiency Scale

Your level of facilitation can be characterized on a scale from *Novice* to *Highly Accomplished*. In *Lemons to Lemonade: Resolving Problems in Meetings, Workshops, and PLCs*, Garmston and Zimmerman (2013) describe these levels. Use their rubric to characterize your level of facilitation. They offer suggestions for next steps at each level.

Facilitator Stage	Facilitator Characteristics	Group Response	What You Can Do
Unaware	Lacks knowledge or information about facilitation or intervention skills. Passively accepts what happens in meetings as outside of her or his control. Attributes problems to others, not to her or his leadership of the meeting.	Groups respond with frustration and report that meetings waste time, overwhelm them, have unproductive conflict, and often spin endlessly on topics of little value.	• Acquire essential foundation knowledge of facilitation work by reading the seminal book *How to Make Meetings Work* by Doyle and Straus (1976), *Unlocking Group Potential for Improving Schools* by Garmston (2012), *Manager's Guide to Effective Meetings* by Streibel (2003), or *Best Practices for Facilitation* by Sibbet (2002).
Novice	Knows basic facilitation skills—how to get a group's attention, set focus and agenda, and manage transitions. May have difficulty leading decision-making processes. Sees the difficult participant as an impediment to progress and lacks skills to intervene effectively.	Meeting tones are not consistent; sometimes the work goes well and other times it is stalled. This inconsistent positive reinforcement may give the illusion that the group is more capable; however, when things get tough, the meeting breaks down. Groups blame a difficult person for the problem and are not aware of any contribution they, as a group, might be making to problems. When the difficult person is absent, everyone notices how well the meeting went.	• Volunteer to facilitate portions of meetings to automate basic facilitation moves. • Practice facilitation principles and moves when working with students. • Observe colleagues. Take notes about their decisions and explore their thinking after the session. • Cofacilitate and have a reflecting conversation afterward. • Facilitate and seek coaching. • Learn more about problem-solving in groups and intervening by reading books such as *Unlocking Group Potential* by Garmston (2012), *Don't Just Do Something, Stand There! Ten Principles for Leading Meetings That Matter* by Weisbord and Janoff (2007), or *The Leader's Handbook: A Guide to Inspiring Your People and Managing the Daily Workflow* by Scholtes (1998).

(Continued)

(Continued)

Facilitator Stage	Facilitator Characteristics	Group Response	What You Can Do
Proficient	Has basic facilitation skills and can manage routine problems effortlessly in meetings. Views exceptional problems as challenges to solve over time. After a meeting, reflects and learns by mentally revising the possible interventions and outcomes. Considers multiple options to employ should behaviors happen again.	Groups perceive their meetings as effortless and may not attribute the success to the facilitator. However, when the facilitator is absent they begin to notice a qualitative difference. A strong facilitator can become paternalistic—keeping order, but not helping group members grow and learn. Groups can become dependent on the leader and stuck in their growth.	• View *Focusing Four* video to observe a master facilitator conducting a consensus session (Garmston & Dolcemascolo, 2009). • Schedule a planning conversation with a colleague prior to a difficult meeting and reflect with him or her after the session. • Seek every opportunity to practice and schedule a planning conversation with a colleague before the meeting and reflect with this person after the meeting. • Become a facilitative participant in meetings you attend. This means you practice these skills when not the formally appointed leader. • Seek new references that have skill-building information and read and envision how to apply the skills. Find an opportunity to practice. • Create a quick reference library with books such as *Thinker Toys: A Handbook of Business Creativity* by Michalko (1991), *The Presenter's Fieldbook: A Practical Guide* by Garmston (2005), or *Resonate: Present Visual Stories That Transform Audiences* by Duarte (2010).

Facilitator Stage	Facilitator Characteristics	Group Response	What You Can Do
Accomplished	Is able to respond, adapt, and improvise in the face of uncertainty. Sees self as responsible to the group's success and does not blame others. Consciously works to shift responsibility for facilitation and intervention principles to the group. Able to teach facilitation skills and interventions.	Groups report that they learn not only about how to do their job better but also how to work effectively with others. They begin to appreciate the quiet voice that finally speaks up or the loud voice that shows humility. They understand how dissenting views can be catalysts for deeper thinking. They transfer facilitation and intervention skills to other aspects of their life. Skilled facilitators quietly celebrate when they observe explicit carryover of skills used in one setting to another. For example, a teacher might use paraphrasing as a way to help students hold onto ideas; a PLC member may use outcome thinking to keep the group focused.	• Learn from a master by reading books such as *The Skilled Facilitator* by Schwartz (2002). • Set specific goals for yourself, such as using certain strategies should an opportunity arise or paraphrasing before taking new comments. • Seek out colleagues with similar skill sets and collaborate on ideas. • Keep a facilitator's notebook with ideas, references, and reflections. • Join an online community or follow a blog on organizational development. See the list created by Terrence Seamon at http://learningvoyager.blogspot.com/2006/12/od-blogs-abound.html. • Seek opportunities to teach about facilitation and interventions.
Expert	Acts intuitively. Has many sets of linked steps that are performed unconsciously. Conscious of choices being made and could reveal the meta-cognition of facilitation to others. Regularly teaches the group about interventions using graphics, modeling, and third-point teaching.	Groups report that they are also learning to facilitate groups in effective ways. Members are increasingly willing to and capable of assuming leadership positions in this and other groups.	• Learn about how stages of adult development affect decision-making by reading books such as *Immunity to Change: How to Overcome It and Unlock the Potential in Yourself and Your Organization* by Kagan and Lahey (2009). • Teach, observe, and coach others.

Source: Adapted from Doyle, M., & Straus, D. (1976). *How to make meetings work: The new interaction method.* New York, NY: Jove and Garmston, R. J. & Zimmerman, D. P. (2013). *Lemons to Lemonade: Resolving Problems in Meetings, Workshops, and PLCs.* Thousand Oaks, CA: Corwin.

13.3 Sample Facilitation Activity: Appreciative Inquiry Protocol

*Instructions to the Coach: All materials for this activity can be downloaded at **resources.corwin. com/mathematicscoaching**.*

Facilitation focus: Fostering positive relationships and building community
Outcomes: Participants will be able to connect with others in the group to focus on positive activities regarding the teaching and learning of math.
Preparation: • Copy the Appreciative Inquiry protocol (adapted for mathematics) for each participant. • Read Coaching Lessons From the Field to see how one coach used this protocol.
Description of activity: *Note: For more detailed directions, see Tool 13.3 Appreciative Inquiry.* • Distribute one Appreciative Inquiry protocol to each participant. • Arrange participants in pairs to conduct interviews (participants will be asked to share their partner's story in the next stage). • Arrange participant pairs into groups of four or six. • Ask groups to record themes they noticed from the paired-interview discussion on sticky notes. • Ask participants ○ to discuss how the themes might be grouped, organized, and named. ○ to imagine what it would be like if their students were regularly engaged in this type of activity. ○ to design a mathematics classroom in which this might be possible. Record on chart paper. ○ what might need to happen for this to become a reality in their classrooms. • Summarize by highlighting what has emerged from the conversation.

NOTES

_____

_____

_____

_____

_____

_____

_____

_____

_____

_____

_____

_____

_____

_____

_____

_____

_____

For Tool 13.3 Appreciative Inquiry

Instructions to the Coach: This tool can be used as a facilitation strategy to balance participation during a PLC meeting.

AI Stage: *Dream*	AI *Dream* Prompt
➤ Ask teachers to remain in groups of four to six. ➤ Ask teachers to chart, draw, or record the group's ideas.	Changing and improving our students' engagement in mathematics learning (insert specific topic) requires imagination and vision. Let's build on the themes we developed. Think about what it would be like if your students were regularly engaging in (insert specific topic). Imagine what this would look like. Imagine what you would be doing. Imagine what students would be doing. Imagine your classroom filled with rich student (insert specific topic). 1. As you imagine, consider the following: 　• Particular structures in the classroom environment that invite student engagement 　• Mathematical routines 　• Mathematical learning experiences 　• Classroom culture 　• Particular teacher behaviors/actions you want to incorporate 2. Discuss with your group your thoughts and ideas from your imagining. 3. What do you notice about the group's ideas?
AI Stage: *Design*	**AI *Design* Prompt**
➤ Ask teachers remain in groups of four to six. ➤ Make sure teachers have access to chart paper and markers. Some may want additional materials, such as construction paper.	Using the ideas from the group and the presentation on (insert specific topic), design a classroom environment filled with lively mathematical learning. Include as many details as possible. Be ready to share your design.
AI Stage: *Deliver*	**AI *Deliver* Prompt**
➤ Ask teachers to locate their original partners to discuss and reflect. ➤ Ask teachers to share their designs.	Think about your group's design and all of the designs that you heard about from the other groups. With your partner, discuss the following questions: 1. What can you do to bring your idea to a reality? 2. What is the smallest step you can take right now? 3. How can you ensure that you will take this action to deliver your design?

NOTES

_____

_____

_____

_____

_____

_____

_____

13.4 Sample Facilitation Activity: Paraphrasing

Facilitation focus: Clarifying thoughts and intention of participants
Outcomes: Participants will be able to clarify and understand the thinking of the group.
Preparation: • No physical materials required. Facilitator needs to be proficient at paraphrasing (see Chapter 11).
Description of activity: • While paraphrasing is an important coaching skill, it is also very useful (and often neglected) when facilitating meetings like PLCs. In meetings, paraphrasing can help participants connect ideas, understand each other, and move forward in meeting their goals. A skillful facilitator can also use paraphrasing in resolving conflicts (see Garmston & Wellman, 2016, and Garmston & Zimmerman, 2013b, for detailed information about paraphrasing as a facilitation tool).   • Audio recording meetings and scripting your paraphrases can be a valuable growth experience for you as the facilitator. Use the accompanying template to support you.

NOTES

_____
_____
_____
_____
_____
_____
_____
_____
_____
_____
_____
_____
_____
_____
_____
_____
_____
_____
_____
_____
_____
_____
_____

Template for Tool 13.4 Paraphrasing

Instructions to the Coach: In the left-hand column, script the paraphrases from your meeting. In the right-hand column, analyze your paraphrases using these questions to guide you:

- *What are you noticing about the effect of your paraphrases on the participants?*
- *How might you make your paraphrases more concise?*
- *In what ways did your paraphrases support the meeting goals?*
- *How did your paraphrases make the meeting more efficient?*

Meeting _____ **Date** _____

Scripted Paraphrase	Analyzing Your Paraphrases

13.5 Sample Facilitation Activity: Inclusion Strategies

Facilitation focus: Supporting participants in transitioning from their busy lives into the meeting	
Outcomes: Participants will be able to focus and prepare to engage in the meeting.	
Preparation: These sample inclusion activities are from the *Adaptive Schools Sourcebook* (Garmston & Wellman, 2016). Additional details on these strategies can be found at www.thinkingcollaborative.com.	
First Job	Ask each participant to share their first paid job. This activity is a very quick one that can be used to help participants relate to similar activities from their young adulthood.
What's on Your Plate?	Hand each participant an unwaxed, white paper plate and markers. Ask participants to write what is "on their plate" (e.g., what they are dealing with, what is causing stress in their life). Ask each participant to share one idea from their plate. Then, ask all participants to physically take their plate and set it on a table on the other side of the room. Explain that they are to temporarily "set aside" what is on their plate in order to focus on the meeting.
Grounding	Form groups of four to six. Explain that the purpose of the activity is to focus on respectful listening and to honor all voices in the room. Explain that participants will take turns responding and all others will listen. When all have responded to the prompt, the first speaker summarizes for the group. Post the prompt on chart paper so all can see. Possible prompts: <ul><li>What is the commitment that brought you to this meeting?</li><li>My expectations for this meeting are . . .</li><li>My relationship to this topic is . . .</li></ul>

Adapted from: Garmston, R., J. & Wellman, B. M. (2016). *The adaptive school: A sourcebook for developing collaborative groups* (3rd ed.). Lanham, MA: Rowman & Littlefield.

NOTES

_____
_____
_____
_____
_____
_____
_____
_____
_____
_____
_____
_____
_____
_____
_____
_____

13.6 Sample Facilitation Activity: Task Talk Protocol

Facilitation focus: Balancing participation
Outcomes: Participants will be able to explore different facets of lesson planning and implementation. The facilitator will be able to balance participation through the use of a protocol.
Preparation: • Copy one set of the Task Talk Placemat on the next page and Task Talk Cards on the subsequent pages for the PLC. If it is a large PLC, you may want to divide into smaller groups. • All materials can also be downloaded at resources.corwin.com/mathematicscoaching.
Description of activity: The Task Talk protocol can be used to support teachers in exploring different facets of lesson planning and implementation of a mathematics task. This protocol is adapted from the concept of writing workshops, where teachers pose questions, offer suggestions, and collaborate over one task. Ask PLC participants to bring a mathematics task they plan to teach to the next PLC meeting. • Ask participants to share their tasks. The group then selects one task to use with the Task Talk protocol. • Have everyone solve the selected task. • Place the task cards face down in each of the four sections of the Task Talk Placemat. • In round-robin fashion, have a participant select a card from any section of the placemat, read it aloud to the group, and begin a conversation based on the selected mathematics task. • Complete several rounds (as time allows). Document decisions agreed upon regarding the task. • Summarize by highlighting what has emerged from the conversation.

Task Talk Protocol Source: Kobett et al. (2015) Adapted from Garmston, R., J. & Wellman, B. M. (2016). *The adaptive school: A sourcebook for developing collaborative groups* (3rd ed.). Lanham, MA: Rowman & Littlefield.

NOTES

_____

_____

_____

_____

_____

_____

_____

_____

_____

_____

_____

_____

_____

_____

_____

_____

_____

_____

_____

_____

Placemat for Tool 13.6 Sample Facilitation Activity: Task Talk Protocol

Student	Teacher

Task Design	Bringing It Together

Task Talk Protocol Source: Kobett et al. (2015)

Cards for Tool 13.6 Sample Facilitation Activity: Task Talk Protocol (Kobett et al., 2015)

Student Questions

Select an individual student and imagine how the student will respond to the task.	How do you envision students will work together?
What questions might students pose while working on the task?	How might you support student collaboration?
How might you support students in exhibiting MP 3 (Critique the reasoning of others)?	In what ways might you differentiate the task for differing student populations?
What should the students learn from the task?	What do you anticipate students might do?
What questions might the students ask?	Which Math Practices should students exhibit?
What misconceptions might a student have?	What vocabulary will the students need to know or develop for the task?

Cards for Tool 13.6 Sample Facilitation Activity: Task Talk Protocol (Kobett et al., 2015)

Teacher Questions

How will the teacher facilitate student collaboration? What specific teacher moves might you observe?	What facilitating questions will be used to open the lesson?
What does the teacher look like and sound like during this lesson?	How does the teacher establish an environment for students that signifies respect and rapport?
How might you support students in exhibiting MP 3 (Critique the reasoning of others)?	What type of environment must the teacher develop for students to engage in this task? What does this look like?
How might the teacher communicate expectations for reasoning, thinking, and collaborating while problem-solving?	What formative assessment techniques could the teacher use during the lesson? At what point in the lesson might this happen?
How might the teacher manage classroom behavior during this lesson? What does this look like?	How will the teacher engage the students in the learning (so that the students are equally as engaged as the teacher)?
How might the teacher flexibly respond to student understanding during the lesson? What might this look like?	In what ways might the teacher communicate with families about teaching rich tasks in the classroom?

Cards for Tool 13.6 Sample Facilitation Activity: Task Talk Protocol (Kobett et al., 2015)

Task Design

What standards might connect to this task?	How might you launch the task?
What materials or tools could you use to support student learning?	What might be some options in grouping students to encourage collaboration and problem-solving?
What questions might you ask while students are working?	How can you connect this task to other content areas?
How might you motivate a struggling learner?	How might you transition from the launch part of the lesson to student collaboration?
How might you extend this lesson if a group finishes before other groups are done?	How might the arts connect to this task?
How might you support students persevering through the task?	In what ways might you differentiate the task for differing student populations?

Bringing It All Together

How might the teacher close the task? What does this look like?	Describe the lesson by working backwards from closure to launch.
What explicit connections should be made from the task to mathematical understanding?	What are some things that could go wrong during the closure of this lesson? How might the teacher respond?
How will the teacher ensure the students understand the point of the task?	What happens tomorrow to connect to the closure?
How will the teacher select groups to share? In what order? Why?	How might the teacher build vocabulary during the closure?
What are three things that will happen in the closure of this task?	How will the teacher ensure the students understand the point of the task?
How will the teacher record key ideas while closing the lesson?	What happens tomorrow? How will you connect this lesson to tomorrow's lesson?

Appendix

Bookmarks

online resources You can download the bookmarks at
resources.corwin.com/mathematicscoaching.

As you have read throughout this book, improving mathematics learning hinges on a strong focus on the eight Mathematical Practices and the eight *Shifts in Classroom Practices* (and related Teaching Practices). Both the Mathematical Practices and the Teaching Practices are *big* ideas. The bookmarks offer more concrete elements of those big ideas, helping teachers to make sense of what these Practices might look like in the classroom.

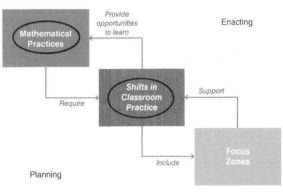

Ideas for making the bookmarks:

- Copy on regular paper and laminate. This way is long lasting, and teachers can write on the laminated bookmark and then wipe clean. Also, it has the texture of a bookmark you might buy in a bookstore!
- Copy on bright, unusually-colored cardstock. This is also long lasting, and easy to find in a stack of papers on a desk.

Ideas for using the bookmarks:

- For professional learning, the options are endless. We have described some of our favorite ideas in Tools 12.8, 12.9, 12.10, and 12.11.
- For your coaching cycle, ask the teacher to review the bookmark and think about which *Shifts* and/or Mathematics Practices will be the focus of a lesson. You can mark this on *your* bookmark, and they can mark it on their bookmark. The reflecting conversations can revisit the selected Practices and/or *Shifts*.
- In lesson study, the bookmarks can be critical to the planning process, ensuring that the lesson attends to selected Mathematics or Teaching Practices.
- In teaching, encourage teachers to keep the Mathematics Practice bookmark on a clipboard, or otherwise nearby. As they are observing students, they can look for and compliment students on the Practices they witness. Such comments communicate that these practices are important to the teacher and are, in fact, what it means to be "good at" mathematics.

    For students! With teachers, make a giant bookmark of the Mathematical Practices as a poster. Encourage teachers to regularly ask students which practices they used in a lesson, and which practices they saw their peers demonstrating.

Teaching Practices & Shifts in Classroom Practice Look Fors

Learning Goals

Shift 1: States a standard
→ Communicates learning expectations

- ☐ Goals are appropriate, challenging, and attainable.
- ☐ Goals are specific to the lesson and clear to students.
- ☐ Goals connect to other mathematics.
- ☐ Goals are revisited throughout the lesson.

Tasks

Shift 2: Routine tasks
→ Reasoning tasks

- ☐ Uses engaging, high-cognitive-demand tasks with multiple solution pathways.
- ☐ Uses tasks that arise from home, community, and society.
- ☐ Uses how, why, and when questions to prompt students to reflect on their reasoning.

Representations

Shift 3: About representations
→ Through representations

- ☐ Uses tasks that lend themselves to multiple representations.
- ☐ Selects representations that bring new mathematical insights.
- ☐ Gives students time to select, use, and compare representations.
- ☐ Connects representations to mathematics concepts.

Mathematical Discourse

Shift 4: Show-and-tell
→ Share-and-compare

- ☐ Helps students share, listen, honor, and critique each other's ideas.
- ☐ Helps students consider and discuss each other's thinking.
- ☐ Strategically sequences and uses student responses to highlight mathematical ideas and language.

Purposeful Questions

Shift 5: Questions seek expected answers
→ Questions illuminate and deepen student understanding

- ☐ Questions make the mathematics visible.
- ☐ Questions solidify and extend student thinking.
- ☐ Questions elicit student comparison of ideas and strategies.
- ☐ Strategies are used to ensure every child is thinking of answers.

Procedural Fluency

Shift 6: Replicating procedures
→ Selecting efficient strategies

- ☐ Gives students time to think about different ways to approach a problem.
- ☐ Encourages students to use their own strategies and methods.
- ☐ Asks students to compare different methods.
- ☐ Asks why a strategy is a good choice.

Productive Struggle

Shift 7: Mathematics-made-easy
→ Mathematics-takes-time

- ☐ Employs ample wait time.
- ☐ Talks about the value of making multiple attempts and persistence.
- ☐ Facilitates discussions on mathematical error(s), misconception(s), or struggle(s) and how to overcome them.

Evidence of Student Thinking

Shift 8: Valuing correct answers
→ Valuing students' thinking

- ☐ Identifies strategies or representations that are important to look for as evidence of student understanding.
- ☐ Makes just-in-time decisions based on observations, student responses to questions, and written work.
- ☐ Uses questions or prompts that probe, scaffold, or extend students' understanding.

www.corwin.com/math

Everything You Need for Mathematics Coaching: Tools, Plans, and a Process That Works for Any Instructional Leader by Maggie B. McGatha and Jennifer M. Bay-Williams with Beth McCord Kobett and Jonathan O. Wray. Thousand Oaks, CA: Corwin, **www.corwin.com/math**. Copyright © 2018 by Corwin. All rights reserved. Reproduction authorized only for the local school site or nonprofit organization that has purchased this book.

online resources

Available for download at **resources.corwin.com /mathematicscoaching.** Copyright © 2018 by Corwin.

www.corwin.com/math

Mathematical Practices & Student Look Fors

1. Make sense of problems and persevere in solving them.

- ☐ Analyze information (givens, constraints, relationships).
- ☐ Make conjectures and plan a solution pathway.
- ☐ Use objects, drawings, and diagrams to solve problems.
- ☐ Monitor progress and change course as necessary.
- ☐ Check answers to problems and ask, "Does this make sense?"

2. Reason abstractly and quantitatively.

- ☐ Make sense of quantities and relationships in problem situations.
- ☐ Create a coherent representation of a problem.
- ☐ Translate from contextualized to generalized or vice versa.
- ☐ Flexibly use properties of operations.

3. Construct viable arguments and critique the reasoning of others.

- ☐ Make conjectures and use counterexamples to build a logical progression of statements to support ideas.
- ☐ Use definitions and previously established results.
- ☐ Listen to or read the arguments of others.
- ☐ Ask probing questions to other students.

4. Model with mathematics.

- ☐ Determine equation that represents a situation.
- ☐ Illustrate mathematical relationships using diagrams, two-way tables, graphs, flowcharts, and formulas.
- ☐ Check to see whether an answer makes sense within the context of a situation and change a model when necessary.

5. Use appropriate tools strategically.

- ☐ Choose tools that are appropriate for the task (e.g., manipulative, calculator, digital technology, ruler).
- ☐ Use technological tools to visualize the results of assumptions, explore consequences, and compare predictions with data.
- ☐ Identify relevant external math resources (digital content on a website) and use them to pose or solve problems.

6. Attend to precision.

- ☐ Communicate precisely, using appropriate terminology.
- ☐ Specify units of measure and provide accurate labels on graphs.
- ☐ Express numerical answers with appropriate degree of precision.
- ☐ Provide carefully formulated explanations.

7. Look for and make use of structure.

- ☐ Notice patterns or structure, recognizing that quantities can be represented in different ways.
- ☐ Use knowledge of properties to efficiently solve problems.
- ☐ View complicated quantities both as single objects and as compositions of several objects.

8. Look for and express regularity in repeated reasoning.

- ☐ Notice repeated calculations and look for general methods and shortcuts.
- ☐ Maintain oversight of the process while attending to the details.
- ☐ Evaluate reasonableness of intermediate and final results.

www.corwin.com/math

Everything You Need for Mathematics Coaching: Tools, Plans, and a Process That Works for Any Instructional Leader by Maggie B. McGatha and Jennifer M. Bay-Williams with Beth McCord Kobett and Jonathan O. Wray. Thousand Oaks, CA: Corwin, **www.corwin.com/math**. Copyright © 2018 by Corwin. All rights reserved. Reproduction authorized only for the local school site or nonprofit organization that has purchased this book.

Available for download at
resources.corwin.com /mathematicscoaching.
Copyright © 2018 by Corwin.

www.corwin.com/math

Glossary

Active engagement: A description of the high level of attention, curiosity, interest, optimism, and passion that a person shows when she or he is learning. See also **student engagement**.

Affective conflict: Disagreements over personalized anger or resentment.

Algorithm: A designated procedure or process for a particular purpose.

Appreciative Inquiry interview: A conversation with a teacher, fostering positive relationships and building community.

Artifact: As it relates to teaching, an actual piece of student work or recorded content.

Authentic context: A situation or story that is familiar to the lived experiences of a student and that is an actual context in which the selected mathematics would be used.

Autobiographical listening: The desire to interject your personal thoughts and ideas into a conversation while listening.

Break-apart: Breaking a number down into smaller parts, typically relative to place value, to make it easier to use the number in computation/problem-solving.

Case: A vignette or video with a designated professional learning purpose.

Classroom discourse: Meaningful conversations in classrooms that can include asking questions, listening to students, or encouraging students to listen, respond, and explain their thinking.

Clinical interview: See **diagnostic interview**.

Coaching: As it relates to instruction, it means providing support to other teachers focused on increasing self-directedness and improving learning opportunities for students. See also **mathematics coaching**.

Coaching cycle: A process of planning, gathering data, and reflecting that focuses on a particular goal for teaching and/or learning.

Conceptual understanding: Comprehension of mathematical concepts, operations, and relations.

Cognitive conflict: Disagreements about important differences of opinion.

Cognitively demanding: Requiring thinking at a higher level, requiring more than retrieval of facts or demonstrating known procedures.

Content knowledge: Understanding related to topics within a domain, such as mathematics. See also **mathematical knowledge for teaching**.

Cooperative learning: A form of small-group activity in which structures are put in place to ensure students are interdependent and accountable.

CRA (concrete–representational–abstract): An instructional progression that begins with ideas that are concrete, visible, and familiar, gradually moving toward mathematical representations, and eventually to abstract mathematics concepts, with the goal of connecting concrete concepts to abstract mathematical ideas.

Culturally responsive mathematics instruction (CRMI): Teaching that focuses on the big ideas of mathematics, makes that content relevant to the learner, attends to a student's identity, and ensures a student's contributions are valued by his or her peers and by the teacher.

Diagnostic assessment: See **diagnostic interview**.

Diagnostic interview: A form of data gathering from a student that helps to discover what specific understandings and skills she or he has related to a concept, giving the teacher insights into how to support that student's learning. See also **clinical interview**.

Differentiating instruction: Designing a lesson in such a way that each student, with individual knowledge bases and learning needs, has access to the content and is able to meaningfully engage in the content.

Differentiating professional development: Designing professional development in such a way that each teacher, with individual knowledge bases and learning needs, has access to the content and is able to meaningfully engage in the professional development.

Discourse: See **classroom discourse**.

Effective Teaching Practices: When capitalized, this phrase refers to the eight specific teaching practices described in *Principles to Actions: Ensuring Mathematical Success for All;* when not capitalized, it refers to teaching practices, in general, that are connected to supporting student learning.

Efficiency: The combination of selecting a strategy that has the potential to reach a solution quickly and the skill at being able to navigate that selected strategy relatively quickly.

Emergent multilingual student: A student who is learning English as he or she is learning content. The term replaces *English language learner* as a more accurate description of these students.

Engagement technique: A technique teachers can infuse into their teaching to support high participation in their students.

Exit task: A fully developed rich task or problem that assesses the learning intention for the day, a cluster of learning intentions, or a standard.

Explicit strategy instruction: A phrase that is defined in different ways, but used in this book to mean teaching in which concepts, procedures, and connections between concepts and procedures receive explicit attention.

Facilitator: Person who creates an environment in which teachers feel safe to take risks and collaborate and who directs the processes used in a meeting to support trusting relationships in order to meet group goals.

Feedback: Questions, prompts, and statements provided to students about their learning as part of the formative assessment process.

Five Strands of Mathematical Proficiency: From the National Research Council's book *Adding It Up: Helping Children Learn Mathematics*, the descriptions of what mathematically proficient people demonstrate as they engage in doing mathematics. See also **mathematical proficiency**.

Fluency: Having or showing mastery of a subject or skill; navigating a problem or situation with ease.

Focus Zone: A term used within this book intended to describe a selected subset of teaching skills, such as analyzing student work.

Focusing: A questioning pattern in which the teacher asks questions based on the students' thinking to support students in thinking at high levels.

Formative assessment: The process of monitoring student learning to gather evidence to provide feedback and adjust instruction.

Formative assessment plan: A comprehensive assessment plan that indicates the kinds of formative techniques students will use, when they will use them, and how they will use them.

Formative 5: Five assessment techniques as described by Fennell, Kobett, and Wray: interviews, observations, Show Me, hinge questions, and exit tasks.

Funneling: A questioning pattern in which the teacher leads the student through a series of questions to the teacher's desired end.

High-level: An adjective describing tasks or questions that require cognition above knowledge level (e.g., on Bloom's taxonomy).

Hinge question: A question asked at a particular point during a lesson that informs the teacher's next instructional step.

Instructional strategy: A way that a teacher organizes and engages students, such as pairing them with peers or presenting work using a gallery walk.

Interview: A formative assessment strategy in which teachers ask purposeful questions to collect evidence of a student's mathematical understanding.

Inquisitive listening: Becoming overly curious about the details of the person's story that might not be important to the conversation.

IRF (initiation–response–feedback): A pattern of questioning in which the teacher asks a question, a student responds, and the teacher provides feedback.

Learning partners: An active engagement strategy for pairing students.

Lesson cycle: A three-step process that includes planning, data gathering, and reflecting on a lesson.

Look for: A term used to describe the actual visible example of a broader idea, such as a mathematical practice or teaching practice.

Low-level: An adjective describing tasks or questions that require cognition limited to knowledge level (e.g., on Bloom's taxonomy).

Mathematical knowledge for teaching (MKT): A phrase used to describe the additional knowledge that is necessary to teach mathematics—beyond what is needed to do mathematics.

Mathematical practice: The description of what a student/person does when he or she is mathematically proficient. In this book, we use this term to refer to any of the eight Mathematical Practices described in the *Common Core State Standards for Mathematics* (NGA Center & CCSSO, 2010; see Table 1.1).

Mathematical proficiency: Five intertwining strands of mathematics: conceptual understanding, procedural proficiency, strategic competence, adaptive reasoning, and productive disposition (NRC, 2001).

Mathematically proficient student: A student who exhibits the five mathematical proficiencies or the eight Mathematical Practices.

Mathematics coaching: Providing support to teachers of mathematics related to increasing self-directedness and improving mathematics teaching and related opportunities for students to learn mathematics.

Multiple entry points: A descriptor used for tasks that can be approached in a variety of ways.

Observation: As used in this book, a formative assessment strategy in which teachers watch students carefully to determine their mathematical understanding.

Open-ended: A term used to describe tasks or questions that have numerous ways to engage or respond.

Parallel task: A form of tiered task, in which students have a *choice* of which task they will solve.

Paraphrase: A concise rephrasing of another's words to clarify understanding.

Peer-assisted learning: Learning from another student in the classroom in an intentionally designed manner (i.e., from how a task is structured or students are grouped).

Pencast: A digital audio and visual recording of notes.

Presenter: A person who leads an activity, supporting the development of knowledge and skills that teachers can apply in their practice.

Procedural fluency: The ability to select and apply procedures accurately, efficiently, and flexibly; to transfer procedures to different problems and contexts; to build or modify procedures from other procedures; and to recognize when one strategy or procedure is more appropriate to apply than another.

Procedure: A process. In mathematics, it refers to a process for reaching the solution to a problem.

Professional development: An event (can take many forms) in which a group of teachers convene for focused and structured learning.

Professional learning: What teachers and other school professionals learn; the intended result of a professional development experience.

Professional learning community (PLC): A group of people that is collectively working together to solve a problem of practice.

Protocol: As used in to this book, a predetermined procedure—for example, for examining student work.

Rapport: Making both verbal and nonverbal connections with another person.

Rubric: A measure that indicates the criteria for success.

Scaffolding: A term used in lesson design that involves adding more structure to the task in order to support student thinking.

Self-assessment: A tool that is used to reflect on one's own knowledge, skills, or beliefs.

***Shift in Classroom Practice*:** A term used within this book to describe a continuum of possible teaching actions related to effective teaching. Eight *Shifts* are described within this book (see Figure 1.2).

Show Me: A formative assessment strategy in which teachers ask students to demonstrate mathematical understanding through mathematical representations.

Solution listening: Listening in order to offer a solution for the other person's problem.

Solution strategy: A way that a person decides to apply mathematical knowledge and solve a problem.

Strategy: A process that is flexible, as compared to an algorithm, which is a set series of steps. See also **instructional strategy** and **solution strategy**.

Student engagement: A description of the level of attention, curiosity, interest, optimism, and passion that students show when they are learning. See also **active engagement**.

Support function: A term used to refer to different leadership roles and ways of supporting others, such as coaching, consulting, collaborating, and evaluating (Costa and Garmston, 2016).

Talk move: A specific way a teacher asks a question or waits on a response in order to elicit student participation or support student understanding.

Task: A problem or exercise designed for students; a set or series of problems designed for students.

Taxonomy: A classification or grouping. Bloom's Taxonomy is a classification of levels of cognition, from lower level to higher level.

Teaching practice: An action a teacher does as part of helping students learn;

in this book, it refers to eight specific Effective Teaching Practices, as described in *Principles to Actions* (NCTM, 2014). See also Table 2.1.

Think-aloud: A learning strategy in which the learner is asked to say aloud what she or he is thinking during problem-solving.

Thinking strategies: As used in this book, specific teacher moves, such as think-pair-share, that are used in order to provide time for students to think and share their thinking.

Tiered task: A problem that has been adapted to offer several options, varying the numbers or the structure of the problem, but with the same learning outcomes.

Total Participation Technique (TPT): The term used in Himmele and Himmele's book *Total Participation Techniques: Making Every Student an Active Learner* to describe instructional strategies that engage students and provide time for them to think and share their thinking.

TPT Cognitive Engagement Model: The model described in Himmele and Himmele's *Total Participation Techniques: Making Every Student an Active Learner* that delineates cognition and participation continua.

Vignette: Short story of an event, often used as an example or exemplar.

Wait time: The amount of time between when a teacher finishes asking a question and the time in which a student is asked to respond; a talk move that focuses on providing more time between a question being posed and requesting a response.

Whole-class discussion: Asking questions to an entire class of students and facilitating so that students listen to their peers and contribute to the conversation. See also **discourse**.

Worthwhile task: A phrase commonly used to mean a task that focuses on important mathematics and engages students in higher-level thinking, among other features.

References

Aguilar, E. (2016). *The art of coaching teams: Building resilient communities that transform schools*. San Francisco, CA: Jossey-Bass.

Aguirre, J. M., Mayfield-Ingram, K., & Martin, D. (2013). *The impact of identity in K–8 mathematics learning and teaching: Rethinking equity-based practices*. Reston, VA: National Council of Teachers of Mathematics (NCTM).

Anderson, L. W., & Krathwohl, D. R., (Eds.). (2001). *A taxonomy for learning, teaching and assessing: A revision of Bloom's Taxonomy of educational objectives: Complete edition*. New York, NY: Addison Wesley Longman.

Baker, S., Lesaux, N., Jayanthi, M., Dimino, J., Proctor, C. P., Morris, J.,… & Newman-Gonchar, R., (2014). *Teaching academic content and literacy to English learners in elementary and middle school* (NCEE 2014-4012). Washington, DC: National Center for Education Evaluation and Regional Assistance (NCEE), Institute of Education Sciences, U.S. Department of Education. Retrieved from the NCEE website: http://ies.ed.gov/ncee/wwc/publications_reviews.aspx.

Barkley, S. G. (2010). *Quality teaching in a culture of coaching* (2nd ed.). Lanham, MD: Rowman & Littlefield.

Bartell, T., Wager, A., Edwards, A., Battey, D., Foote, M., & Spencer, J. (2017). Toward a framework for research linking equitable teaching with the Standards for Mathematical Practice. *Journal for Research in Mathematics Education*, *48*(1), 7–21.

Bay-Williams, J., McGatha, M., Kobett, B., & Wray, J. (2014). *Mathematics coaching: Resources and tools for coaches and leaders, K–12*. New York, NY: Pearson Education.

Bay-Williams, J. M., & Stokes Levine, A. (2017). The role of concepts and procedures in developing fluency. In D. Spangler & J. Wanko (Eds.), *Enhancing professional practice with research behind principles to actions*. Reston, VA: NCTM.

Billings, E. M. H. (2017). SMP that help foster algebraic thinking. *Teaching Children Mathematics*, *23*(8), 476–483.

Black, P., & Wiliam, D. (1998). Inside the black box: Raising standards through classroom assessment. *Phi Delta Kappan*, *80*(2), 139–148.

Bloom, B. S., Engelhart, M. D., Furst, E. J., Hill, W. H., & Krathwohl, D. R. (1956). *Taxonomy of educational objectives: The classification of educational goals. Handbook I: Cognitive domain*. New York, NY: Longman, Green.

Boaler, J. (2016). *Mathematical mindsets*. San Francisco, CA: Jossey-Bass.

Brodesky, A. R., Fagan, E. R., Tobey, C. R., & Hirsch, L. (2016). Moving beyond one-size-fits-all PD: A model for differentiating professional learning for teachers. *Journal of Mathematics Education Leadership*, *17*(1), 20–37.

Bryk, A. S., & Schneider, B. (2002). *Trust in schools: A core resource for school improvement*. New York, NY: Russell Sage Foundation.

Carr, J. F., Herman, N., & Harris, D. E. (2005). *Creating dynamic schools through mentoring, coaching, and collaboration*. Alexandria, VA: Association for Supervision and Curriculum Development (ASCD).

Cassone, J. D. (2009). Differentiating mathematics by using task difficulty. In D. Y. White & J. S. Spitzer (Eds.), *Mathematics for every student: Responding to diversity, grades pre-K–5* (pp. 89–98). Reston, VA: NCTM.

Celedón-Pattichis, S. (2009). What does that mean? Drawing on Latino and Latina students' language and culture to make mathematical meaning. In M. W. Ellis (Ed.), *Responding to diversity: Grades 6–8* (pp. 59–74). Reston, VA: NCTM.

Chapin, S., O'Conner, C., & Anderson, N. C. (2013). *Talk moves: A teacher's guide for using classroom discussions in math* (3rd ed.). Sausalito, CA: Math Solutions.

Colton, A. B., & Langer, G. M. (2005). Looking at student work. *Educational Leadership*, *62*(5), 22.

Cooperrider, D. L., & Whitney, D. (2005). *Appreciative inquiry: A positive revolution in change*. San Francisco, CA: Berrett-Koehler.

Costa, A. L., & Garmston, R. J. (2016). *Cognitive coaching: Developing self-directed leaders and learners* (3rd ed.). Lanham, MA: Rowman & Littlefield.

Covey, S. R. (2008). *The speed of trust: The one thing that changes everything*. New York, NY: Simon & Schuster.

Covey, S. R. (2013). *The 7 habits of highly effective people: 25th anniversary edition*. New York, NY: Simon & Schuster.

Crespo, S. (2002, April). *Doing mathematics and analyzing student work: Problem-solving discourse as a professional learning experience*. Paper presented at the American Educational Research Association, New Orleans.

Crockett, M. D. (2002). Inquiry as professional development: Creating dilemmas through teachers' work. *Teaching and Teacher Education*, *18*(5), 609–624.

Darling-Hammond, L., Hyler, M. E., & Gardner, M. (2017). *Effective teacher professional development*. Palo Alto, CA: Learning Policy Institute.

Doyle, M., & Straus, D. (1976). *How to make meetings work: The new interaction method*. New York, NY: Jove.

Dunleavy, T. K. (2015). Delegating mathematical authority as a means to strive toward equity. *Journal of Urban Mathematics Education*, *8*(1), 62–82.

Ellison, J., & Hayes, C. (2009). Cognitive coaching. In J. Knight (Ed.), *Coaching: Approaches and perspectives*. Thousand Oaks, CA: Corwin.

English, L. D., Fox, J. L., & Watters, J. J. (2005). Problem posing and solving with mathematical modeling. *Teaching Children Mathematics, 12*(3), 156.

Erikson, K. I., Hillman, C. H., & Kramer, A. F. (2015). Physical activity, brain, and cognition. *Current Opinion in Behavioral Sciences, 4,* 27–32.

Fan, L., & Bokhove, C. (2014). Rethinking the role of algorithms in school mathematics: A conceptual model with focus on cognitive development. *ZDM, 46*(3), 481–492.

Fennell, F., Kobett, B. M., & Wray, J. A. (2017). *The formative 5: Everyday assessment techniques for every math classroom.* Thousand Oaks, CA: Corwin.

Flores, M., Hinton, V., & Strozier, S. (2014). Teaching subtraction and multiplication with regrouping using the concrete–representational–abstract sequence and strategic instruction model. *Learning Disabilities Research and Practice, 29*(2), 75–88.

Fontana, D., & Fernandes, M. (1994). Improvements in mathematics performance as a consequence of self-assessment in Portuguese primary school pupils. *British Journal of Educational Psychology, 64*(3), 407–417.

Garmston, R. J., & Wellman, B. M. (2016). *The adaptive school: A sourcebook for developing collaborative groups* (3rd ed.). Lanham, MA: Rowman & Littlefield.

Garmston, R. J., & Zimmerman, D. P. (2013a). The collaborative compact: Operating principles lay the groundwork for successful group work. *Journal of Staff Development, 34*(2), 10–16.

Garmston, R. J., & Zimmerman, D. P. (2013b). *Lemons to lemonade: Resolving problems in meetings, workshops, and PLCs.* Thousand Oaks, CA: Corwin.

Gersten, R., Beckmann, S., Clarke, B., Foegen, A., Marsh, L., Star, J. R., & Witzel, B. (2009). *Assisting students struggling with mathematics: Response to intervention (RtI) for elementary and middle schools* (NCEE 2009–4060). Washington, DC: National Center for Education Evaluation and Regional Assistance.

Gibbons, L. K., Lewis, R. M., & Batista, L. N. (2016). The sandwich strategy: No matter how you slice it, analyzing student work together improves math instruction. *Journal of Staff Development 37*(3), 14–18, 37.

Ginsburg, H. (1997). *Entering the child's mind: The clinical interview in psychological research and practice.* New York, NY: Cambridge University Press.

Ginsburg, H. P. (2009). The challenge of formative assessment in mathematics education: Children's minds, teachers' minds. *Human Development, 52*(2), 109–128.

Glaser, J. E. (2014). *Conversational intelligence: How great leaders build trust and get extraordinary results.* New York, NY: Bibliomotion.

Glickman, C. D., Gordon, S. P., & Ross-Gordon, J. M. (2001). *Supervision and instructional leadership: A developmental approach.* Boston, MA: Allyn & Bacon.

Goldhammer, R. (1969). *Clinical supervision: Special methods for the supervision of teachers.* New York, NY: Holt, Rinehart and Winston.

Goleman, D. (2006). *Social intelligence: The new science of human relationships.* New York, NY: Random House.

Gómez, C. L. (2010). Teaching with cognates. *Teaching Children Mathematics, 16*(8), 470–474.

Grinder, M. (1993). *ENVoY: Your personal guide to classroom management* (2nd ed.). Battle Ground, WA: Michael Grinder & Associates.

Guskey, T. R. (2000). *Evaluating professional development.* Thousand Oaks, CA: Corwin.

Guskey, T. R. (2017). Where do you want to get to? Effective professional learning begins with a clear destination in mind. *The Learning Professional, 38*(2), 32–37.

Haas, E., & Gort, M. (2009). Demanding more: Legal standards and best practices for English language learners. *Bilingual Research Journal, 32,* 115–135.

Hattie, J., Fisher, D., Frey, N., Gojak, L. M., Moore, S. D., & Mellman, W. (2016). *Visible learning for mathematics, grades K–12: What works best to optimize student learning.* Thousand Oaks, CA: Corwin.

Hattie, J. A. (2009). *Visible learning: A synthesis of 800+ meta-analyses on achievement.* Abingdon, UK: Routledge.

Herbel-Eisenmann, B. A., & Breyfogle, M. L. (2005). Questioning our patterns of questioning. *Mathematics Teaching in the Middle School, 10*(9), 484–489.

Hiebert, J., & Grouws, D. A. (2007). *Effective teaching for the development of skill and conceptual understanding of number: What is most effective?* Research brief for NCTM. Reston, VA: NCTM. Retrieved from http://www.nctm.org/news/content.aspx?id=8448.

Hiebert, J., & Wearne, D. (1993). Instructional tasks, classroom discourse, and students' learning in second-grade arithmetic. *American Educational Research Journal, 30,* 393–425.

Hill, H. C., Rowan, B., & Ball, D. L. (2005). Effects of teachers' mathematical knowledge for teaching on student mathematical achievement. *American Education Research Journal, 42*(2), 371–406.

Himmele, P., & Himmele, W. (2017). *Total participation techniques: Making every student an active learner.* Alexandria, VA: ASCD.

Hodges, T. E., Rose, T. D., & Hicks, A. D. (2012). Interviews as RtI tools. *Teaching Children Mathematics, 19*(1), 30–36.

Hufferd-Ackles, K., Fuson, K. C., & Sherin, M. G. (2015). Describing levels and components of a math-talk learning community. In E. A. Silver & P. A. Kenney (Eds.), *More lessons learned from research, Volume 1: Useful and usable research related to core mathematical practices* (pp. 125–134). Reston, VA: NCTM.

Huinker, D., & Bill, V. (2017). *Taking action: Implementing effective mathematics teaching practices in K–grade 5.* Reston, VA: NCTM.

Iacoboni, M. (2008). *Mirroring people: The science or empathy and how we connect with others.* New York, NY: Picador.

Isaacs, W. (1999). *Dialogue and the art of thinking together.* New York, NY: Doubleday.

Jacobs, V. R., Ambrose, R. C., Clement, L., & Brown, D. (2006). Using teacher-produced videotapes of student interviews as discussion catalysts. *Teaching Children Mathematics, 12*(2), 276–281.

Jensen, E. (2008). *Brain-based learning: The new paradigm of teaching*. Thousand Oaks, CA: Corwin.

Joyce, B., & Calhoun, E. (2016). What are we learning about how we learn? *Journal of Staff Development, 37*(3), 42–44.

Joyce, B., & Showers, B. (2003). *Student achievement through staff development*. Alexandria, VA: ASCD.

Kazemi, E., & Franke, M. L. (2004). Teacher learning in mathematics: Using student work to promote collective inquiry. *Journal of Mathematics Teacher Education, 7*(3), 203–235.

Kee, K., Anderson, K., Dearing, V., & Shuster, F. (2017). *Results coaching next steps: Leading for growth and change*. Thousand Oaks, CA: Corwin.

Keene, E. O. (2014). All the time they need. *Educational Leadership, 72*(3), 66–71.

Killerman, S., & Bolger, M. (2016). *Unlocking the magic of facilitation*. Austin, TX: Impetus Books.

Killion, J. (2013). *School-based professional learning for implementing the Common Core: Unit 2, facilitating learning teams*. Oxford, OH: National Staff Development Council (Learning Forward).

Kingore, B. (2006, Winter). Tiered instruction: Beginning the process. *Teaching for High Potential*, 5–6. Retrieved from http://www.bertiekingore.com/tieredinstruct.htm.

Knowles, M. (2015). *The adult learner: The definitive classic in adult education and human resource development* (8th ed.). New York, NY: Routledge.

Kobett, B., Stavish, M. Z., Arminio, J., Bertram, A., Frankenford, J., Meskill, A.,… Zickefoose, K. (2015, September). *Build your own task talk community*. Presentation at the Association for Maryland Mathematics Educators Early Career Conference.

Krathwohl, D. R. (2002). A revision of Bloom's Taxonomy: An overview. *Theory Into Practice, 41*(4), 212–218.

Levin, T., & Long, R. (1981). *Effective instruction*. Washington, DC: ASCD.

Lewis, K. (2014). Difference not deficit: Reconceptualizing mathematical learning disabilities. *Journal for Research in Mathematics Education, 45*(3), 351–396.

Lieberman, M. D. (2014). *Social: Why our brains are wired to connect*. Oxford, UK: Oxford University Press.

Lipton, L., Wellman, B., & Humbard, C. (2003). *Mentoring matters: A practical guide to learning-focused relationships*. Sherman, CT: MiraVia.

Little, J. W., Gearhart, M., Curry, M., & Kafka, J. (2003). Looking at student work for teacher learning, teacher community, and school reform. *Phi Delta Kappan, 85*(3), 185–192.

Mancl, D. B., Miller, S. P., & Kennedy, M. (2012). Using the concrete-representational–abstract sequence with integrated strategy instruction to teach subtraction with regrouping to students with learning disabilities. *Learning Disabilities Research and Practice, 27*(4), 152–166.

Mateas, V. (2016). Debunking the myths about the Standards for Mathematical Practice. *Mathematics Teaching in the Middle School, 22*(2), 92–99.

McGatha, M., & Darcy, P. (2010). Rubrics at play. *Mathematics Teaching in the Middle School, 15*(6), 328–336.

McMaster, K. L., & Fuchs, D. (2016). Classwide intervention using peer-assisted learning strategies. In S. Jimerson, M. Burns, & A. VanDerHeyden (Eds.), *Handbook of response to intervention* (pp. 253–268). New York, NY: Springer.

Mid-continent Research for Education and Learning (McREL). (2009). *Looking at the classroom from the other side*. Retrieved from http://mcrel.typepad.com/mcrel_blog/2009/05/looking-at-the-classroom-from-the-other-side.html.

Moschkovich, J. N. (2009, March). *How do students use two languages when learning mathematics? Using two languages during conversations*. NCTM Research Clip. Reston, VA: NCTM.

Nasir, N.S., Cabana, C., Shreve, B., Woodbury, E., & Louie, N. (Eds.). (2014). *Mathematics for equity: A framework for successful practice*. New York, NY: Teachers College Press.

National Center for Education Statistics (NCES). (2003). *Teaching mathematics in seven countries: Results from the TIMSS video study*. Washington, DC: U.S. Department of Education.

National Council of Teachers of Mathematics (NCTM). (2007). *Research brief: Effective strategies for teaching students with difficulties in mathematics*. Retrieved from www.nctm.org/news/content.aspx?id=8452.

National Council of Teachers of Mathematics (NCTM). (2014). *Principles to actions: Ensuring mathematical success for all*. Reston, VA: Author.

National Governors Association Center for Best Practices & Council of Chief State School Officers (NGA & CCSSO). (2010). *Common Core State Standards for Mathematics*. Washington, DC: Authors.

National Research Council (NRC). (2001). *Adding it up: Helping children learn mathematics*. J. Kilpatrick, J. Swafford, & B. Findell (Eds.). Mathematics Learning Study Committee, Center for Education, Division of Behavioral and Social Sciences and Education. Washington, DC: National Academy Press.

PBS Teacherline. *Developing mathematical thinking with effective questions*. Retrieved March 2012 at http://www.pbs.org/teacherline/

Reiss, K. (2015). *Leadership coaching for educators: Bringing out the best in school administrators* (2nd ed.). Thousand Oaks, CA: Corwin.

Renkl, A. (2014). Learning from worked examples: How to prepare students for meaningful problem solving. In V. Benassi, C. E. Overson, & C. M. Hakala (Eds.), *Applying science of learning in education: Infusing psychological science into the curriculum* (pp. 118–130). Retrieved from http://teachpsych.org/ebooks/asle2014/index.php.

Roake, J. (2013). Planning for processing time yields deeper learning. *Education Update, 55*(8), 1, 6–7.

Rowe, M. B. (1972). *Wait-time and rewards as instructional variables, their influence in language, logic, and fate control*. Paper presented at the National Association for Research in Science Teaching, Chicago, IL.

Shepard, L. A. (2008). Formative assessment: Caveat emptor. In C. A. Dwyer (Ed.), *The future of assessment: Shaping teaching and learning* (pp. 279–303). New York, NY: Erlbaum.

Small, M. (2017). *Good questions: Great ways to differentiate mathematics instruction in the standards-based classroom* (3rd ed.). Reston, VA: NCTM.

Small, M., & Lin, A. (2010). *More good questions: Great ways to differentiate secondary mathematics instruction*. Reston, VA: NCTM.

Smith, M., & Stein, M. (2011). *Five practices for orchestrating productive mathematics discussions*. Reston, VA: NCTM.

Smith, M. S., & Stein, M. K. (1998). Selecting and creating mathematical tasks: From research to practice. *Mathematics Teaching in the Middle School, 3*(5), 344–350.

Smith, N. (2017a). *Every math learner: A doable approach to teaching and learning with differences in mind, grades K–5*. Thousand Oaks, CA: Corwin.

Smith, N. (2017b). *Every math learner: A doable approach to teaching and learning with differences in mind, grades 6–12*. Thousand Oaks, CA: Corwin.

Sousa, D., & Tomlinson, C. (2011). *Differentiation and the brain: How neuroscience supports the learner-friendly classroom*. Bloomington, IN: Solution Tree Press.

Spangler, D. A., & Wanko, J. J. (Eds.). (2017). *Enhancing classroom practice with research behind principles to actions*. Reston, VA: NCTM.

Stahl, R. J. (1994). Using "wait-time" and "think-time" skillfully in the classroom. *ERIC Digest* (ED370885).

Star, J. R. (2005). Reconceptualizing procedural knowledge. *Journal for Research in Mathematics Education, 36*(5), 404–411.

Star, J. R., & Verschaffel, L. (2017). Providing support for student sense making: Recommendations from cognitive science for the teaching of mathematics. In Jinfa Cai (Ed.), *Compendium for research in mathematics education*. Reston, VA: NCTM.

Stein, M. K., & Lane, S. (1996). Instructional tasks and the development of student capacity to think and reason: An analysis of the relationship between teaching and learning in a reform mathematics project. *Educational Research and Evaluation, 2*, 50–80.

Stein, M. K., & Smith, M. S. (1998). Mathematical tasks as a framework for reflection: From research to practice. *Mathematics Teaching in the Middle School, 3*(4), 268–275.

Stiggins, R. (2005). From formative assessment to assessment for learning: A path to success in standards-based schools. *Phi Delta Kappan, 87*(4), 324–328.

Stigler, J. W., & Hiebert, J. (2009). *The teaching gap: Best ideas from the world's teachers for improving education in the classroom*. New York, NY: Free Press.

Strauss, V. (2016, September 13). Why some schools are sending kids out for recess four times a day. *The Washington Post*. Retrieved August 30, 2017, from https://www.washingtonpost.com/news/answer-sheet/wp/2016/09/13/recess-four-times-a-day-why-some-schools-are-now-letting-kids-play-an-hour-a-day/?utm_term=.17203ceec291

Tomaz, V. S., & David, M. M. (2015). How students' everyday situations modify classroom mathematical activity: The case of water consumption. *Journal for Research in Mathematics Education, 46*(4), 455–496.

Tomlinson, C. (1999). Mapping a route towards differentiated instruction. *Educational Leadership, 57*(1), 12–16.

Tomlinson, C. A. (2001). *How to differentiate instruction in mixed-ability classrooms* (2nd ed.). Alexandria, VA: ASCD.

Tomlinson, C. A., & McTighe, J. (2006). *Integrating differentiated instruction*. Alexandria, VA: ASCD.

Tschannen-Moran, B., & Tschannen-Moran, M. (2010). *Evocative coaching*. San Francisco, CA: Jossey-Bass.

Tschannen-Moran, M. (2014). *Trust matters: Leadership for successful schools* (3rd ed.). San Francisco, CA: Jossey-Bass.

Van de Walle, J. A., Bay-Williams, J. M., Lovin, L. H., & Karp, K. S. (2018). *Teaching student-centered mathematics: Grades 6–8* (3rd ed.). New York, NY: Pearson Education.

Van de Walle, J. A., Karp, K. S., & Bay-Williams, J. M. (2019). *Elementary and middle school mathematics methods: Teaching developmentally* (10th ed.). New York, NY: Pearson Education.

Wiliam, D. (2007). *Five "key strategies" for effective formative assessment*. Reston, VA: NCTM. Retrieved from http://www.nctm.org/clipsandbriefs.aspx.

Wiliam, D. (2011). *Embedded formative assessment*. Bloomington, IN: Solution Tree.

Wiliam, D., & Leahy, S. (2015). Embedding formative assessment: Practical techniques for K–12 classrooms. West Palm Beach, FL: Learning Sciences International.

Woleck, K. (2010). *Moments in mathematics coaching: Improving K–5 instruction*. Thousand Oaks, CA: Corwin.

Wright, R. J., & Ellemor-Collins, D. L. (2008). Assessing student thinking about arithmetic: Videotaped interviews. *Teaching Children Mathematics, 15*(2), 106–111.

Index

NOTES

NOTES

NOTES

NOTES

NOTES

NOTES

NOTES

Supporting Teachers, Empowering Learners

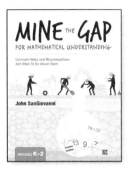 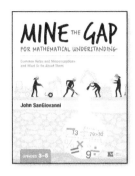 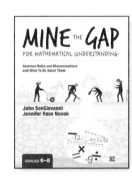

See what's going on in your students' minds, plus get access to hundreds of rich tasks to use in instruction or assessment

John SanGiovanni, Jennifer Rose Novak

Grades K–2, ISBN: 978-1-5063-3768-5
Grades 3–5, ISBN: 978-1-5063-3767-8
Grades 6–8, ISBN: 978-1-5063-7982-1

The what, when, and how of teaching practices that evidence shows work best for student learning in mathematics

John Hattie, Douglas Fisher, Nancy Frey, Linda M. Gojak, Sara Delano Moore, William Mellman

Grades K–12, ISBN: 978-1-5063-6294-6

Move the needle on math instruction with these 5 assessment techniques

Francis (Skip) Fennell, Beth McCord Kobett, Jonathan A. Wray

Grades K–8, ISBN: 978-1-5063-3750-0

Differentiation that shifts your instruction and boosts ALL student learning

Nanci N. Smith

Grades K–5, ISBN: 978-1-5063-4073-9
Grades 6–12, ISBN: 978-1-5063-4074-6

Everything you need to promote mathematical thinking and learning

Page Keeley, Cheryl Rose Tobey

Grades K–12
Volume 1, ISBN: 978-1-4129-6812-6
Volume 2, ISBN: 978-1-5063-1139-5

CM CORWIN MATHEMATICS

N17C44

A SAGE Publishing Company

CORWIN HAS ONE MISSION: to enhance education through intentional professional learning.

We build long-term relationships with our authors, educators, clients, and associations who partner with us to develop and continuously improve the best evidence-based practices that establish and support lifelong learning.

NCSM is a mathematics education leadership organization that equips and empowers a diverse education community to engage in leadership that supports, sustains, and inspires high quality mathematics teaching and learning every day for each and every learner.

NATIONAL COUNCIL OF
TEACHERS OF MATHEMATICS

The National Council of Teachers of Mathematics is the public voice of mathematics education, supporting teachers to ensure equitable mathematics learning of the highest quality for each and every student through vision, leadership, professional development, and research.